P9-BZI-533

Alice and the Space Telescope

MALCOLM LONGAIR

Alice and

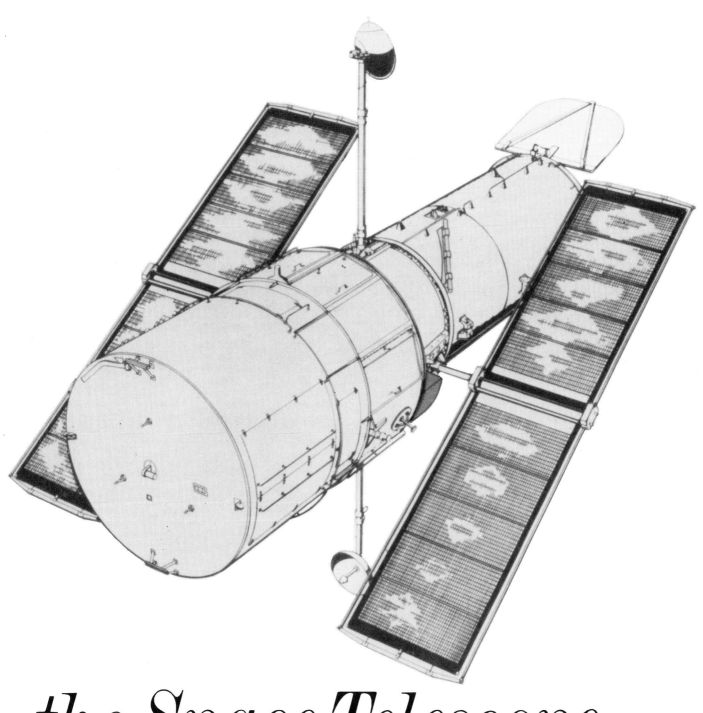

the Space Telescope

THE JOHNS HOPKINS UNIVERSITY PRESS

BALTIMORE AND LONDON

For Mark and Sarah

Published for the Space Telescope Science Institute
(operated by the Association of Universities for Research in Astronomy, Inc.)
by The Johns Hopkins University Press.

©1989
Association of Universities for Research in Astronomy, Inc.,
Baltimore, Maryland
All rights reserved
Printed in the United States of America

The Johns Hopkins University Press
701 West 40th Street
Baltimore, Maryland 21211
The Johns Hopkins Press Ltd., London

The paper used in this publication
meets the minimum requirements of
American National Standard for Information Sciences—
Permanence of Paper for Printed Library Materials
ANSI Z39.48-1984.

Library of Congress Cataloging-in-Publication Data
Longair, M. S., 1941–
Alice and the space telescope.

Bibliography: p. 193
Includes index.
1. Hubble Space Telescope.   I. Title.
QB500.268.L66   1989   522′.29   85-23924
ISBN 0-8018-2831-7 (alk. paper)

Contents

Foreword

The launch of the Hubble Space Telescope in late 1989 will provide optical astronomers with an unprecedentedly clear view of the heavens. The telescope will enable us to distinguish details of celestial objects ten times more finely than is possible from the ground; to observe faint point sources fifty times weaker than can now be discerned; and to extend observation from the ultraviolet to the near infrared without the obscuring effects of the Earth's atmosphere. The expected scientific returns from such improved observational capabilities promise to be extremely significant for all areas of astronomy, from planetary research to cosmology. Understandably, astronomers the world over are looking forward with great excitement to making use of the Hubble Space Telescope.

The Hubble Space Telescope project is a joint enterprise of the National Aeronautics and Space Administration (NASA) and of the European Space Agency (ESA). The telescope will be the most complex, as well as one of the largest, scientific instruments carried into orbit by the Shuttle Transportation System. It is also one of the most technologically demanding space missions yet undertaken and, not unreasonably, one of the most expensive. These features of the program have tended to add to its scientific appeal the spice of space adventure, publicity, and controversy concerning the magnitude of the effort. As a consequence, both legislative and advisory bodies of the United States Government and the general public have become more interested in this mission than is usual for a purely scientific mission.

NASA officials have responded to the unusual challenges and opportunities of the Hubble Space Telescope program in a variety of ways. For instance, on the advice of the National Academy of Sciences, they have entrusted the scientific conduct of the mission to an independent institute, the Space Telescope Science Institute (STScI), which is administered by the twenty members of the Association of Universities for Research in Astronomy, Inc. (AURA). The STScI is responsible for planning the science program, reviewing proposals from the astronomical community, selecting observations, developing an integrated observing plan, receiving the data, performing the necessary calibrations, and archiving and distributing them.

The creation of such an institute is a first for NASA. It appears to be an excellent approach for a mission of such long duration: the Hubble Space Telescope will be serviced and maintained in orbit for twenty years.

Attention has also been given to the question of how the enormous amount of interest in the Hubble Space Telescope and its findings can be satisfied. NASA, ESA, and the STScI are eager to share with the public the wonder and excitement of frontier science, and they will do as much as possible to achieve this goal. The institute itself has been charged with communicating the newly acquired scientific knowledge to other scientists by means of seminars, symposia, and publications, and to the general public via long-term educational programs, planetarium shows, publications, and lectures.

It is within this general framework that we conceived of an annual public lecture sponsored by the STScI and directed to the interested public in Baltimore, as well as to faculty and students of the Johns Hopkins University, on whose campus we reside. The general plan is to invite distinguished scholars to lecture on astronomical subjects related to the Hubble Space Telescope which will have scientific as well as popular appeal. The lectures will be supported by private money raised by the STScI Associates Program.

We were indeed fortunate in having as the first lecturer in this series Malcolm Longair, Astronomer Royal for Scotland and Director of the Royal Observatory, Edinburgh. Professor Longair has served as Interdisciplinary Scientist on the Hubble Space Telescope program since its inception, and he is fully knowledgeable about the Hubble Space Telescope instrumentation and the scientific goals of the program.

He chose to present the Hubble Space Telescope project to the public through the eyes of Lewis Carroll's Alice, and this approach resulted in an hour of sheer delight for his audience. The lecture was given at the Johns Hopkins University campus on the occasion of the 164th meeting of the American Astronomical Society, in Baltimore, on June 12, 1984. The response of the audience, a mixed crowd of astronomers and lay people, was so favorable that STScI and the Johns Hopkins University Press decided to cooperate in the publication of an expanded version of the lecture. The book gives a charming, humorous, but always exact, explanation of what the Hubble Space Telescope is and what it does. It can be read on many levels, and I am sure that it will engage the general reader as well as the professional astronomer.

We owe a great debt of gratitude to Malcolm Longair for his brilliant achievement. He has generously donated his rights to this book to the STScI, and the proceeds will be devoted to a continuing program of public education. We are also grateful to Mr. Jacques T. Schlenger, Chairman of the STScI

Associates Program, and to the Jacob and Annita France Foundation, Mr. Willard Hackerman, and Mrs. Ryda Levi, whose generous contributions made this publication possible. The Hubble Lecture itself is funded each year by a community service grant from the Marion and Henry Knott Foundation of Baltimore.

RICCARDO GIACCONI, Director
Space Telescope Science Institute
Baltimore

IN MEMORIAM

On January 28, 1986, the Space Shuttle *Challenger* was destroyed just seventy-three seconds after lift-off, resulting in the deaths of seven astronauts. This dreadful human tragedy brought home to us all the inherent danger of manned space flight and the dedicated courage of the teams of astronauts involved in the Space Shuttle program. As a result of the accident, the NASA program of shuttle launches was interrupted so that modifications and further safety features could be incorporated into the Shuttle Transportation System. One consequence of this delay was that the launch of the Hubble Space Telescope was postponed. After much effort by NASA to improve shuttle safety, the program of shuttle launches has now resumed, and the expected launch date of the Space Telescope is December 11, 1989.

This book will therefore be published more than five years after the Alice lecture was delivered in Baltimore. It had reached an advanced stage of preparation for publication at the time of the *Challenger* tragedy. We decided that, although astronomy has continued to advance rapidly in the intervening period, the book should be published as a record of the original lecture, incorporating only the appropriate changes of dates and references to astronomical discoveries. In preparing the original text for publication, we emphasized major astronomical and physical problems that remain unresolved in 1989; consequently, there is nothing essential we would wish to change. It was our purpose then to enlighten while entertaining a broad audience, and that is our purpose still.

MALCOLM LONGAIR
RICCARDO GIACCONI

Preface

Alice and the Space Telescope must seem two unlikely bedfellows, and I owe readers an explanation for this unusual combination. I was deeply honored to be invited to give the public lecture at the meeting of the American Astronomical Society which was held in Baltimore in June 1984. It is no coincidence that the Space Telescope Science Institute is located on the Homewood campus of the Johns Hopkins University in Baltimore. The subject of the lecture was therefore to be the Space Telescope (or the Hubble Space Telescope, as it is now formally known) and the great scientific opportunities it offers. The problem I faced is exposed in the first paragraph of Chapter 1—there has been so much written about how marvelous the Telescope will be that it was a challenge to find an original way not only of conveying the excitement of the program but also of giving some insight into what astronomers consider to be the most important scientific issues.

It so happened that I had been reading *Alice's Adventures in Wonderland* and *Through the Looking-glass* to our children, Mark and Sarah, in the splendid edition annotated by Martin Gardner (*The Annotated Alice*, Penguin Books, 1980). Once again I was struck by the pure genius of these immortal fantasies, which have no equal in the English language. I also read at about the same time Chandrasehkar's brilliant short book on Eddington, in which he quotes the splendid parody of "The Walrus and the Carpenter" entitled "The Einstein and the Eddington," by W. H. Williams (see *Eddington: The Most Distinguished Astrophysicist of His Time*, Cambridge University Press, 1983). Somehow these different strands must have become entangled in my mind, and the concept of telling the story of the Space Telescope through Lewis Carroll's characters became inevitable.

Any parody of the Alice books is bound to be but a very pale shadow of the originals. The opportunity was, however, irresistible, and I had great fun composing the "Alice text" which follows as a tribute to the Hubble Space Telescope and as an act of homage to the genius of Carroll. My versions of the characters are much gentler than Carroll's and consequently lack the sharp cutting edge of the originals. But my aims are clearly different from his. Besides the purely educational aspects of the wonderland of modern astronomy, it was a splendid opportunity for gently teasing my colleagues about the way they behave as research scientists. It was also an opportunity for exposing some basic home truths about research in general. What one might hesitate as a professional to write about some of the attitudes of ones colleagues seems perfectly acceptable in the mouths of Wonderland characters.

The annotations that follow the Alice text are in reality a sequence of short essays that give more of the background to the astronomy discussed by Alice and her Wonderland and Looking-glass friends. These reflect current thinking about many of the most important problems of modern astronomy. They may act as a springboard for the enthusiast to tackle the many excellent texts now available on modern astronomy and cosmology. The bracketed numbers in Part I, "Alice and the Space Telescope," refer to note numbers in Part II, "Annotations."

The lecture was given in a very crowded basement lecture room of Remsen Hall at the Johns Hopkins University during a heat wave on 12 June 1984. Although I did not realize it at the time of writing, the very first words of the Alice text were remarkably appropriate as an introduction to the lecture.

I owe acknowledgments to a number of people whose help was essential in the preparation of the original lecture and in preparing the expanded version for publication. Brian Hadley and his colleagues of the Photolabs of the Royal Observatory, Edinburgh, did a magnificent job in preparing the slides for the lecture and much of the photographic material in the book. Riccardo Giacconi has been very supportive of the project to publish an expanded version of the lecture in book form. His colleagues at the Space Telescope Science Institute, Sharon Wanglin, Kim Johnson, and Mark Littman, have been particularly helpful and enthusiastic in helping with the arrangements that were made with the Johns Hopkins University Press. The Press has been particularly accommodating in the publishing of what must seem a rather strange academic book for a university press. Of the many people involved in its production at the Press I am especially grateful to Jack Goellner, Anders Richter, Barbara Lamb, Stephen Kraft, and James Johnston. My special thanks are due to the Associates of the Space Telescope Science Institute, who have agreed to underwrite this undertaking.

My final acknowledgments are to my wife, Deborah, and our children, Mark and Sarah, whose love is a constant inspiration. Among the greatest pleasures in preparing this book has been the thought that I could dedicate it to Mark and Sarah.

Edinburgh, April 1985   MALCOLM LONGAIR

PART I *Alice and the Space Telescope*

1

An Old Friend

It was a very hot sunny day and Alice found it difficult to concentrate upon yet another lecture about the Space Telescope. She seemed to have heard so many lectures about how marvelous the telescope would be that she felt she could give a better lecture than this eccentric professor with the Scottish accent who kept waving his hands excitedly.

She looked down at the list of essays she had to prepare and thought how very interesting they were. She felt sure she would get excellent marks for them. The page read as follows:

SPACE TELESCOPE

Write an essay on each of the following subjects:

1

Why will the Hubble Space Telescope revolutionize
mankind's understanding of the Universe?

2

Describe how you would construct a
Space Telescope.

3

What are the most important pieces of
fundamental science which will be studied
by the Hubble Space Telescope?

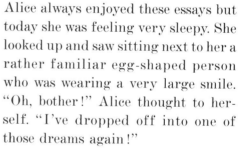

Alice always enjoyed these essays but today she was feeling very sleepy. She looked up and saw sitting next to her a rather familiar egg-shaped person who was wearing a very large smile. "Oh, bother!" Alice thought to herself. "I've dropped off into one of those dreams again!"

I should explain that Alice was now much older than she had been when she first went to Wonderland and Looking-glass Land. Consequently she was now somewhat more cynical than she had been then, but it was always her rule to be polite to people even though she may have been thinking rather less charitable thoughts to herself. I should also explain that Alice was a very good student who had worked hard and diligently at her courses and had achieved very good grades in all her work[1].

But here she was in the lecture theatre, where mysteriously the lecturer and all the other students had suddenly disappeared leaving her with Humpty Dumpty[2], who was looking very well.

"Good morning, Humpty Dumpty," she said. "I am very pleased to see you looking so well after that unfortunate accident when you fell off the wall."

"Oh, it was most unfortunate," replied Humpty Dumpty. "All the king's horses and all the king's men couldn't put me together again. However, they swept up the pieces and sent me to NASA. They're very good at repairing things, you know—*Apollo 13, Solar Max*, eggs[3]. So I am now good as new again."

Alice was very impressed and excited to learn that Humpty Dumpty had been to NASA. "Do you know a lot about the Space Telescope?" she asked.

"I know *everything* about the Space Telescope," said Humpty haughtily. "I know even more than Riccardo Giacconi, Lyman Spitzer, Bob O'Dell, and everyone else associated with the Space Telescope project put together[4]. I can claim to know more about most things than anyone else in the world."

Alice thought to herself, "Sounds like a typical astronomer to me."

"What is that piece of paper you are holding?" asked Humpty.

"It's just the list of essays I have to write about the Space Telescope," Alice replied.

"They look very easy to me," said Humpty. "Let me try the first one : 'Why will the Space Telescope revolutionize mankind's understanding of the Universe?' "

"That *is* a nice topic," said Alice. "I will put the project in its historical perspective and draw analogies with other great scientific projects."

"Nonsense!" said Humpty Dumpty. "The answer is very short. Space Telescope will revolutionize mankind's understanding of the Universe because it cost a billion dollars![5] That's all. They wouldn't have spent the money in the first place if it hadn't been so, now would they?"

"I don't think I will get many marks for that answer," said Alice. She thought to herself, "Maybe professors can get away with remarks like that, but students certainly cannot. I'll have to come back to that question later."

"Let me see the next one," said Humpty. " 'Describe how you would construct a Space Telescope.' " Humpty frowned and looked at the other essay titles. "I would advise," he said pompously, "that you would benefit from the advice of my friends and colleagues in Wonderland and Looking-glass Land. Come along—don't delay."

Somehow, they had traveled very fast to a large building full of scientific instruments, and there were the experts who knew . . .

How to Build a Space Telescope

There were many strange animals running hither and thither, a few borogoves here, some mome raths there, and the odd tove running around with a spanner in its hand[1]. The one thing which struck Alice was that she could not understand the language they were using, although it certainly sounded like English[2].

"The complement of SIs for ST includes, in addition to the WF/PC and FGSs, an FOC, an FOS, an HRS, and an HSP."

"The signal from the FGS passes through the SIC and DH and the SSM, and is transmitted via the TDRSS to the NASCOM and hence to the POCC."

"The software for first-cut data reduction by GOs is provided by the IDTs and by the SDAS project at the STScI."

Alice's head was spinning and so she asked a friendly looking tove, "Could you tell me, please, what these people just said?"

The tove looked at Alice with incomprehension and said in a slow voice, "I cannot interface with your output because of interpersonal no-goes vis-à-vis verbal interaction."

Alice thought for a moment and then asked politely, "Did you just say 'I didn't understand you'?"

The tove thought hard and said, "I read you."

Alice saw that communication with these animals was going to be very difficult. Humpty came to her aid. He banged a nearby scientific instrument with a hammer and said loudly, "Silence, please! This young person wishes to know how to construct a Space Telescope. Will you please be so kind as to enlighten her?"

There was a pause as this was translated and then all the animals sang in chorus the song "Jabberwocky II." (I give here the translation kindly provided by Humpty Dumpty.)[3]

JABBERWOCKY II

'Twas brillig and the slithy toves
Brought plans of telescopes fair to see.
The Jabberwock, he clapped his hands
And said, "That's just for me."[4]

"Build me a telescope, large and true,
That into space will soar.
2.4 meters in size will do,
The Shuttle cannot handle more.[5]

"A Cassegrain telescope it will be,
Of Ritchey-Chrétien form.
Everyone builds them that way now—
Depart not from the norm![6]

"Diffraction optics, if you please,
With images, sharp as can be.
Ten times sharper than from the Earth
And better in the UV.[7]

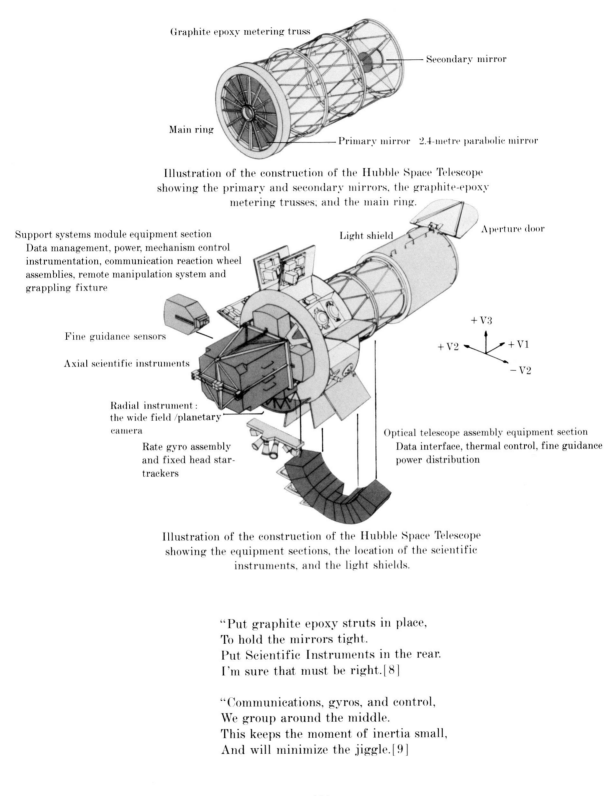

Graphite epoxy metering truss

Secondary mirror

Main ring

Primary mirror   2.4-metre parabolic mirror

Illustration of the construction of the Hubble Space Telescope
showing the primary and secondary mirrors, the graphite-epoxy
metering trusses, and the main ring.

Support systems module equipment section
    Data management, power, mechanism control
    instrumentation, communication reaction wheel
    assemblies, remote manipulation system and
    grappling fixture

Light shield

Aperture door

Fine guidance sensors

Axial scientific instruments

+ V3

+ V2   + V1

— V2

Radial instrument:
the wide field /planetary
camera

Rate gyro assembly
and fixed head star-
trackers

Optical telescope assembly equipment section
    Data interface, thermal control, fine guidance
    power distribution

Illustration of the construction of the Hubble Space Telescope
showing the equipment sections, the location of the scientific
instruments, and the light shields.

"Put graphite epoxy struts in place,
To hold the mirrors tight.
Put Scientific Instruments in the rear.
I'm sure that must be right.[8]

"Communications, gyros, and control,
We group around the middle.
This keeps the moment of inertia small,
And will minimize the jiggle.[9]

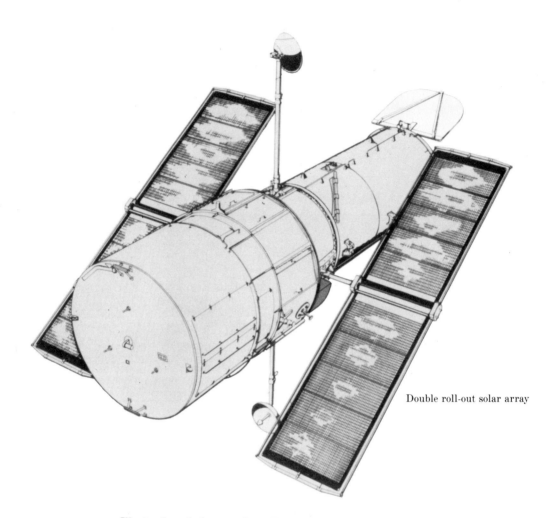

Double roll-out solar array

Illustration of the overall configuration of the Hubble Space
Telescope showing the fore and aft shrouds, the solar array, and
the high-gain antennae.

"Enclose it all in shrouds and vents,
Antennae for radio contact.
All we need is approval now,
And a nice fat juicy contract."[10]

'Twas brillig and the slithy toves,
They rubbed their hands with glee.
"A splendid monument we'll build,
For all mankind to see."

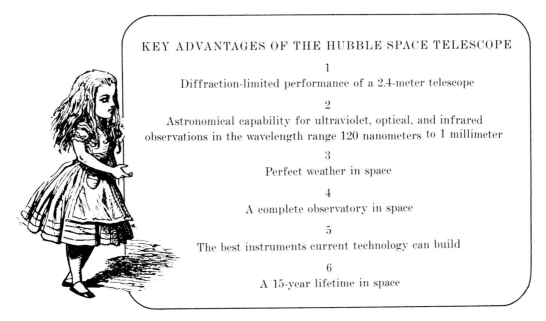

KEY ADVANTAGES OF THE HUBBLE SPACE TELESCOPE

1

Diffraction-limited performance of a 2.4-meter telescope

2

Astronomical capability for ultraviolet, optical, and infrared
observations in the wavelength range 120 nanometers to 1 millimeter

3

Perfect weather in space

4

A complete observatory in space

5

The best instruments current technology can build

6

A 15-year lifetime in space

Alice clapped her hands and all the animals looked very pleased. She went over in her mind the key advantages of the Space Telescope[11].

"Are you sure the telescope will be able to do all these wonderful things?" she asked.

"Yes!" they shouted in chorus.

"And when will it come into operation?"

"The bird will fly in 1986 July!" they all shouted, forgetting that they had finished their song[12]. They seemed unanimous and perfectly confident in their enthusiasm. Alice remembered that it was 1984 but didn't say anything.

"What we need now is the science essay number 3," said Humpty Dumpty. In bustled the Red Queen. "Hurry up!" she said, and off they zoomed to Baltimore and the Space Telescope Science Institute.

The Planetary Caucus-race

At the Institute a huge battle was raging in which everyone seemed to be fighting everyone else[1]. Alice was very surprised to see the scientists fighting one another. They were all shouting loudly and one or two of them were being hit repeatedly very hard by all the other animals.

"Whatever is going on here?" asked Alice.

The Red Queen shouted rudely[2], "Can't you recognize a meeting of the American Astronomical Society when you see one?[3] Everyone fights everyone else and everyone hits particularly hard those with whom everyone else disagrees!"

"But I thought these were scientists, seeking the higher truths by learned discussion and rational argument," said Alice.

"Where have you been hiding for the last hundred years?" muttered the Red Queen.

In all the disorder, there was one group which seemed to be keeping itself somewhat apart from all the others. Alice went up closer and saw that there were lots of birds and animals taking part in a Caucus-race. In fact, they were all her old friends from Wonderland who had taken part in the previous Caucus-race[4]. This time the race seemed to be somewhat better organized. Everyone seemed to be running in more-or-less circular paths about one very small mouse who looked rather terrified by the whole proceedings.

Among the animals she recognized her old friend the Mouse, who was running as fast as he could. She called out to him, "Mr. Mouse, can you come and help me, please?"

The Mouse stopped running and at the same time all the others crowded round to see what had stopped the race.

"Oh, it's her again," said the Lory rather rudely. "I hope she remembered better prizes this time! We don't want sweets or candy—we want gold medals and certificates!"

Alice pretended she had not heard him and told the Mouse of the object of her visit.

"We are all Solar System and planetary scientists here," said the Mouse. "Nobody's quite sure whether we are astronomers or not[5]. That's why we keep ourselves to ourselves here."

"I don't think you should worry about that," said Alice. "As long as the science is exciting and the Space Telescope can make observations which will advance our understanding of these subjects, I am only too eager to learn from you. I was pleased to see that you have developed a new version of the Caucus-race."

"We had to," said the Mouse. "We all work on different sorts of objects—moons, planets, asteroids, comets. We couldn't have all these bodies crashing into one another and so now we all run in different orbits. Everyone here agreed that we should all run about the spot marked by my little boy."

Alice knew exactly the response expected of her in Wonderland. "Well, it's only natural," she said, "that all the planets should move in orbits around the son!"

"Welcome back to Wonderland!" they all shouted together.

They were all delighted that Alice understood them so well. "If you can understand that," said the Mouse, "you won't have any problems with planetary science."

"What are you all doing then?" asked Alice.

"It's amazing how the animals have all chosen different topics appropriate to their characters," said the Mouse. "For example, the large birds and animals work on the giant planets—Jupiter, Saturn, Uranus, and Neptune—the small animals work on the inner planets—Mercury, Venus, and Mars—and the smallest animals study the asteroids and other small bodies in the Solar System."

"And what do you do, Mr. Dodo?" asked Alice.

"I work on extinct bodies in the Solar System. They wouldn't let me work on anything else," said the Dodo sadly.

"And what do these birds do?"

"Oh, they are mostly hawks," said the Mouse. "They are not really involved with the Space Telescope. They are just making sure that the Space Telescope project does not take all the money away from other Solar System projects. There are lots of hawks in astronomy too, you know."

Alice thought the hawks looked rather nervous and suspicious. She noticed that they bit the other birds and animals from time to time with their sharp beaks. "Well, I hope they are successful in protecting their projects," said Alice. "The combination of Space Telescope and other planetary missions must be very exciting."

"Exactly so," said the Eaglet, who was the largest bird present. "I work on Jupiter and I want to study its weather patterns. Do you know why the Planetary Camera on the Space Telescope is so called?"[6]

"Yes," said Alice. "With that instrument, the whole of Jupiter will just fit neatly into a single picture. Every picture of Jupiter will have the same sort of detail we can see in some of the beautiful pictures taken by the *Voyager* space probes which flew past Jupiter[7]. Since Jupiter has the largest image of all the planets, this means that each planet can be photographed with the Space Telescope in a single exposure."

"Quite so," said the Eaglet. "The big advantage of Space Telescope is that we can take pictures of a planet like Jupiter every day and observe directly the changing weather patterns over the whole planet. At about the same time that the Space Telescope first flies, NASA will also launch the *Galileo* space

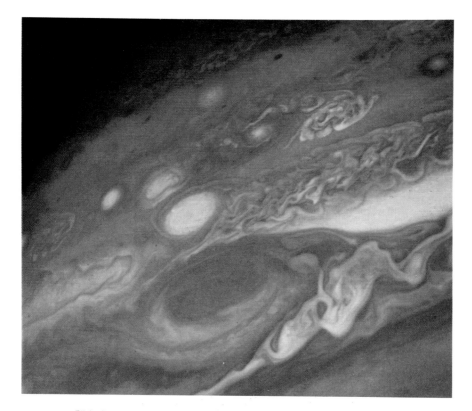

This image of Jupiter was taken by the *Voyager 2* spacecraft in
1979. It shows the typical quality of the images of Jupiter which
will be taken by the Hubble Space Telescope, which will, however,
be able to capture the whole planet in a single exposure.

probe[8], which will go into orbit about Jupiter and fly past a number of its
satellites. *Galileo* will enable planetary scientists to make detailed studies of
Jupiter, its satellites, and the environment of the great planet itself. I will
want to relate the detailed observations made by the *Galileo* satellite to the
large-scale properties of the planet which will be observed by the Space
Telescope. In this way, I hope to be able to understand how the environment
of the planet influences its atmosphere. The same will apply for all the
satellites of Jupiter which will be studied by *Galileo*. This is a particularly
exciting study because the major satellites of Jupiter all seem to be so
different. They range from completely inert objects like Callisto, which
seems to have a heavily cratered surface similar to that of our own Moon, to
exotic active satellites like Io"[9].

"This must be especially exciting for Io, which is the closest of the four
major satellites to Jupiter," said Alice.

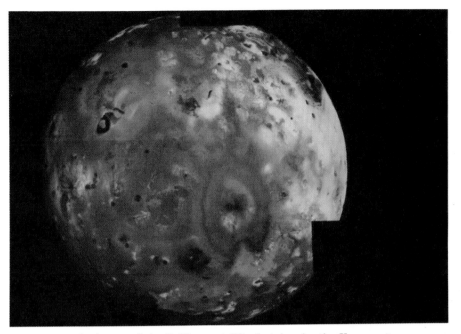

An image of the Galilean satellite Io taken by the *Voyager 2*
spacecraft. Its yellow-orange-red appearance is due to the various
forms of sulphur and sulphur compounds on its surface.

A volcanic eruption on Io observed by the *Voyager 2* spacecraft
as it passed the satellite.

The rings of Saturn as observed by the *Voyager 2* spacecraft.

"Indeed so," said the Eaglet. "The volcanoes on Io are quite remarkable events, pouring sulphurous material over the surface of Io and blasting it high above the surface. It will be very interesting to look for the effects of these volcanoes when *Galileo* passes close to the satellite. By making observations with the Space Telescope we will be able to monitor the volcanic activity on Io at the same time that *Galileo* makes its observations."

"Don't forget the other outer planets!" said the Lory. "Saturn, Uranus, and Neptune will be revealed in great detail and we will be able to observe their weather patterns over long periods. We don't really know what the atmospheres of these planets are like; they may well have weather systems very different from those we know about from studies of the nearer planets[10]. We will especially want to look at the rings of Saturn to see if they change with time. The fine structure in the rings observed by the *Voyager* space probes was observed for only a short period. Over longer periods of time, 10 to 20 years, during which observations can be made from the Space Telescope, we may be able to see changes in the rings themselves, and this may enable us to understand why the beautiful fine structure in the rings can persist for so long"[11].

"We know that there are rings about three of the major planets—Jupiter, Saturn, and Uranus," said Alice. "Space Telescope may tell us whether or not Neptune possesses a ring system as well. All of them will be accessible from the Space Telescope, even peculiar systems like the rings of Uranus. The rings of Uranus circle its equator, just like the rings of Jupiter and Saturn. The surprise is that the axis about which the planet and its rings rotate lies in the plane of orbit of Uranus about the Sun. The system is lying on its side—it really is a most remarkable system."

"Don't forget about me!" shouted the smallest Crab. "I work on Pluto and I want to see its little moon, Charon. Nobody has really seen Pluto and Charon clearly yet. From Space Telescope pictures we should be able to determine their sizes and shapes directly"[12].

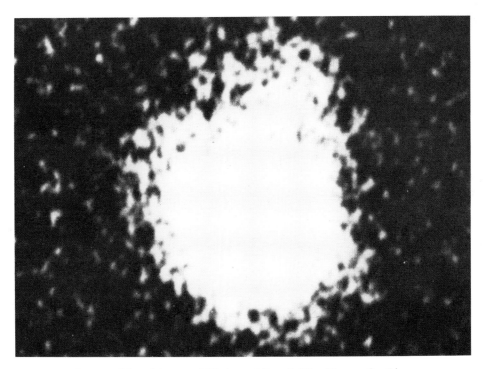

A ground-based image of Pluto and its satellite, Charon. In this photograph the satellite is not properly separated from Pluto, although it is at its most distant point from the planet. It is the "bump" at the top right of the image of Pluto. With the Hubble Space Telescope, Charon will be clearly resolved as a separate object.

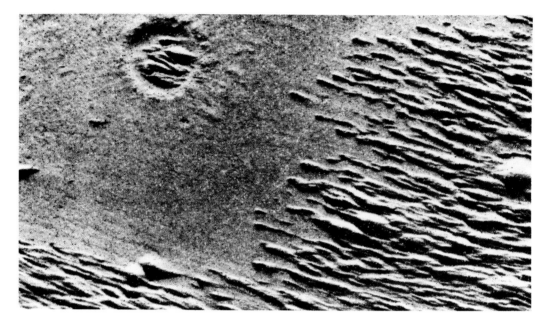

Photographic evidence of dust on Mars. This photograph, taken
by one of the *Viking* orbiters, shows dunes formed by wind
erosion.

"I'm sure you will observe it with Space Telescope," said Alice kindly.

"Now, behave yourself!" said his mother, who was a very large red Crab. "We will all get our turn. I work on Mars and I want to study the terrible dust storms which can blow up and cover large areas of the surface of the planet with dust[13]. It's all highly relevant to our understanding of how our own atmosphere might behave if large amounts of dust were ejected into it for one reason or another."

Alice was beginning to feel that she was hearing rather a lot about weather on other planets.

"This all sounds very exciting," she said, "but do you think planetary meteorology is really a suitable subject for study by an astronomical telescope?"

They all glared at her. The Mouse spoke up for all of them. "You had better take our planetary weather studies very seriously. If we do not understand how planetary atmospheres work, we may end up doing something catastrophic to the Earth's atmosphere, something which could lead to the extinction of all life on Earth, astronomers included"[14].

There was a long pause. Suddenly, the conversation had become deadly serious. Everyone realized that we must understand better the processes which preserve the atmosphere of our own planet, and comparative studies of other planetary atmospheres can provide important clues to the way in which our own atmosphere behaves.

The silence was broken by the Dodo. "Speaking as someone who is already extinct, I would like to draw your attention to primitive bodies in the Solar System. I feel particularly well qualified to speak about dead objects, being, so to speak, a dead object myself. You will remember that small bodies in the Solar System, such as asteroids, meteorites, and comets, are probably made up of the primitive material out of which the planetary system was formed[15]. The surface of the Earth is very different now from what it was like when it was first formed. Then, there were no plants, trees, animals, or human beings, all of which have done so much to change the face of our planet."

A montage of the two satellites of Mars, Phobos *(left)* and Deimos *(right)*, taken by the *Viking 1* orbiter. In this image, the two satellites are shown in their correct relative proportions.

A photograph of Comet Halley taken by the UK Schmidt Telescope
in Australia on 10 March 1986.

"And how will you be able to study the primordial material?" asked Alice.

"With difficulty," said the Dodo, "unless we are able to excite or heat up these bodies somehow."

"That's my field!" said the Mouse. "I work on comets[16]. When they come close to the Sun on their extremely elongated orbits, their surfaces are heated up by sunlight and by the impact of particles blown off the Sun. The volatile gases and dust which are blown off produce the characteristic comet's tail."

"You must be very excited about the reappearance of Halley's Comet in 1986," said Alice.

"Oh dear," said the Mouse. "Mine is a long sad tale!"

"I've heard that before," thought Alice. "I wonder if he has a new version of his poem?"

"Do tell me your tale"[17], she said politely.

The Mouse began sorrowfully.

THE MOUSE'S TALE

"In nineteen eighty-
six, my dear, Halley's
Comet will appear,
We've known it would
for many a year,
But now, alas,
alack, I fear,
We'll miss it
by over a year,
And so I shed
a bitter tear.
The story's
tragic, it
is clear.
Enough to
drive a
Mouse to
beer. No
hope, no
laughter,
and no
cheer.
Place
lilacs
on my
welcome
bier.
Farewell,
cruel
world,
my end
is
near.

Everyone was very moved. It is true that Halley's Comet will pass closest to the earth in April 1986 and that the Space Telescope will be launched well after that date. Once the commissioning of the telescope and its instruments has been completed, the comet will be on its way out of the Solar System to return again only after another 76 years. It really is a piece of rather bad luck.

The little Mouse went up to his father and whispered rather loudly, "Hey, Daddy! That was super. It was worth all the practice, wasn't it? Do you think they will give you more observing time because they feel sorry for you?"

The Mouse aimed a smack at his son but the little fellow dodged quickly out of the way. "Don't give the game away! We have to use every trick of the trade to win more observing time for comets!"

The other animals were clapping loudly. "An excellent performance!" they shouted. They compared it with other performances they had seen. "Almost as good as Martian storms and the aftermath of nuclear explosions!" "Nearly as good as asteroids crashing into the Earth!" "Better than the use of black holes to solve the energy crisis!"[18]

The Mouse bowed, smiling appreciatively. Alice tried to cheer up the Mouse. "Well, you may be able to observe Halley's Comet on its way out from the Sun, and other comets are bound to appear during the lifetime of the Space Telescope." She remembered how intriguing it was to measure the abundances of the different gases and volatile compounds in comets since these provide clues about the prehistory of the Solar System. The chemicals in the comet may be the same as those from which the planets were formed.

Alice did not quite know what to make of it all. She had to confess that her real interest lay in astronomy rather than in Solar System science. But she had to agree that many of the problems being tackled by the planetary scientists are of more direct relevance to the problems of our own planet than virtually all other astronomical problems. They are related to the problems of the origin and evolution of the planetary system and of how life on Earth originated and is maintained. Many of these studies could be beautifully carried out with the Space Telescope.

"Thank you very much," she said politely. "That was most helpful. But I really must be going now. Goodbye!"

"Goodbye and good luck!" the animals and birds all shouted and immediately set off running another Caucus-race. It was only afterwards that Alice realized that they had forgotten all about their prizes. They certainly deserved them this time.

"Twinkle, Twinkle, Little Star"

As Alice was watching the Caucus-race, the Red Queen came up to her. "Well," she said, "have they tried to brainwash you into believing that all Space Telescope observing time should be devoted to the study of our Solar System? I warn you now that all the astronomers you meet will try to persuade you that their subject needs all the observing time on the Space Telescope."

"That doesn't seem so different from ground-based astronomy, does it?" said Alice. "Let's hear what the other astronomers have to say for themselves."

They moved on to another floor of the building, where there were many separate groups busily working on astronomical problems.

"You should ask these animals about the astronomy they want to study with the Space Telescope. They are friendly for the most part, but I warn you that they are all crazy, even by Wonderland standards. I've heard them all before so I will leave you now. Goodbye!" said the Red Queen, and she bustled off at great speed.

On one door there was a large sign which said

Alice hummed to herself,

> Twinkle, twinkle, little star,
> How I wonder what you are!
> Up above the world so high,
> Like a diamond in the sky.

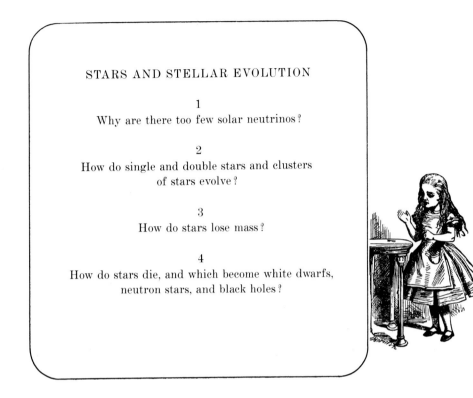

STARS AND STELLAR EVOLUTION

1

Why are there too few solar neutrinos?

2

How do single and double stars and clusters
of stars evolve?

3

How do stars lose mass?

4

How do stars die, and which become white dwarfs,
neutron stars, and black holes?

She thought to herself, "Something very funny is going on here—that poem came out right, which is most unusual for Wonderland. I suppose, however, that nothing could be more appropriate for the study of the stars."

Alice went over in her mind the important questions she wanted to ask about stars and stellar evolution[1]. She thought this was a pretty good list to begin with.

Inside the room she found the Gryphon and the Mock Turtle[2]. They seemed very sad and were reminiscing about the old days. "We used to be the most important thing in astronomy," wept the Mock Turtle. "But now all the bright students want to study black holes, inflationary universes, and quantum gravity."

"That's not true," said Alice. "We must understand stars or else how can we understand everything else in the Universe?"

"How nice to meet a sensible, old-fashioned young lady," said the Gryphon. "If we can hardly understand the Sun, why go to the edge of the Universe to look for problems?"

The Mock Turtle wiped away a tear and said, "Simply because it's there! These silly extragalactic astronomers can see the edge of the Universe and so they want to study it."

"It was so nice in the old days before all these newfangled astronomies—radio, infrared, ultraviolet, and X-ray astronomy—were discovered. No respectable astronomer would touch them with a barge pole—they were all invented by physicists who couldn't tell the difference between a B star and a B movie!" said the Gryphon.

"Quite so," said the Mock Turtle. "In our young days, radio astronomy was what we did on cloudy nights. There was no point in looking through our telescopes and so we listened to the radio instead."

"We did infrared astronomy on frosty nights when we had to wear our thermal underwear to keep warm," sighed the Gryphon.

"We did ultraviolet astronomy after we had been observing in Hawaii. We would lie on the beaches, thinking about astronomy and turning a lovely shade of brown," said the Mock Turtle with a tear in his eye.

"And we did X-ray astronomy when we fell off our telescopes in the dark and broke a leg!" said the Gryphon.

"Ah, happy days!" they sighed together.

"You must agree," said Alice, "that the new astronomies have made some marvelous original contributions to stellar astronomy. Think of the study of stellar winds, which has been revolutionized by ultraviolet observations, or

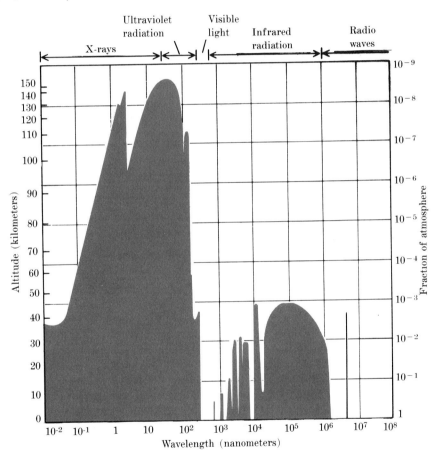

The wavebands which are accessible to astronomical observation at different altitudes. Only the optical, radio, and certain regions of the infrared wavebands are accessible from the ground. X-ray, ultraviolet, and far-infrared astronomy have to be carried out at high altitudes, ideally from telescopes on board space vehicles.

the hot coronae seen around all types of stars in the X-ray waveband, or the X-ray binary stars, or the mass loss rates which have been measured by radio and infrared observations. Isn't that all rather impressive?"

"Reluctantly, we have to agree," said the Gryphon. "But you must understand that, so far as we are concerned, the main effect of these new astronomies has been to take observing time on optical telescopes away from the traditional pursuits of stellar astronomers. I think it is jolly unfair. Why don't the proponents of the new astronomies go and build their own optical telescopes and not use up our observing time?"

"Do you know what happens whenever these people make new discoveries?" said the Mock Turtle. "They immediately want to look at the objects with our optical telescopes to find out what is really there. They cannot even understand what they are looking at without the use of optical telescopes[3]. And yet they think they are the new astronomers."

Alice could see that she had touched upon a large number of very sore points so far as the Gryphon and the Mock Turtle were concerned. She thought it best to change the subject and so she asked, "Well, tell me what worries you most about stellar evolution?"

"The solar neutrinos, or rather the lack of them," said the Gryphon. "There should be three times as many as we observe." Alice remembered that neutrinos are very weakly interacting particles which are produced in large quantities in nuclear reactions in the center of the sun and that they provide one of the very few direct tests of the nuclear processes going on inside stars. They cannot be detected by ordinary telescopes. Huge tanks of cleaning fluid at the bottom of deep mines are needed to detect just a few every year.

There was a very long pause and finally the Mock Turtle burst into a fit of sobs and said sorrowfully, "The Space Telescope is not very good at detecting neutrinos."

By now Alice was getting rather annoyed with these unhappy creatures and so she tried to cheer them up. "Look," she said. "You have the most terrific opportunities to understand all sorts of new important things about stars without bothering about these silly old neutrinos. At least *some* neutrinos have been detected. Think how you would feel if none had been found at all! My professor told me that astronomers never know how to calculate their errors anyway, so let's not get too despondent.

"You must remember that Space Telescope will have the greatest advantage over ground-based telescopes for pointlike objects like stars[4]. Let me tell you some of the exciting things which will become possible[5]:

1. You will be able to determine the relationship between temperature and luminosity for stars in clusters with much greater precision than ever before. This will provide really strong tests of the theory of how stars evolve, which is among the most exact of the astrophysical sciences.

2. You will be able to look into the cores of clusters of stars and identify what types of stars there are there. You will be able to study what is happening dynamically in the very cores of the oldest clusters. No one has been able to do this properly before.

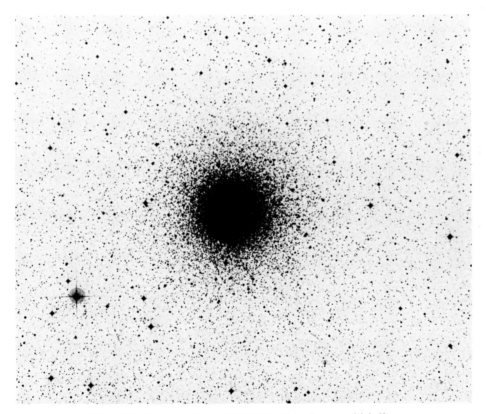

The globular star cluster NGC 362. This cluster, which lies
within our own Galaxy at a distance of about 25,000 light-years,
contains approximately one million stars.

3. You will learn with much greater precision than before how all types of stars lose mass from their outer layers.

4. You will have the capability of measuring the distances of a thousand times as many stars as can be measured from the ground.

[31]

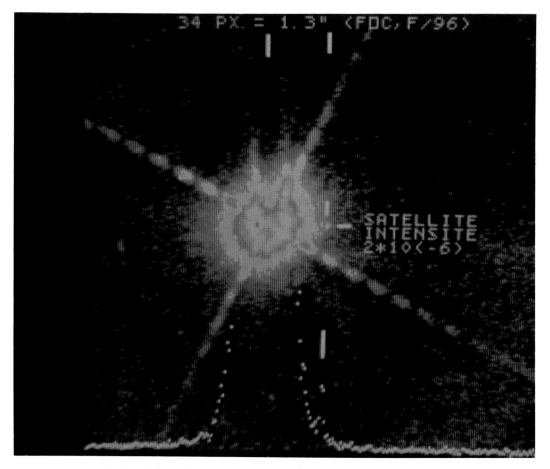

A simulation of the appearance of a Jupiter-like planet close to a
nearby star as it will be observed by the Faint-Object Camera.
The crosslike diffraction spikes on the image of the star are due
to its very great brightness compared to that of the planet.
Although the planet is very faint, being only two-millionths as
bright as the star, its presence can be confirmed by rotating the
telescope and taking another image. The planet, if real, should
appear at a shifted position among the complexities of the
diffraction pattern of the bright star.

5. You will be able to study the very faintest stars that there are in our own
   Galaxy, and, perhaps most exciting of all,

6. You will be able to search for planets around nearby stars.

"Now, doesn't that sound exciting?"
"Yes!" they both said and burst out crying again.

"I've had enough of this," thought Alice, and she slipped out quietly. "I never knew I was so knowledgeable about stars and the Space Telescope. It must be the effect of Wonderland."

Farther along the corridor, Alice came to another door with a large sign on it which said

The door was shut tight but there did seem to be an awful lot of noise coming from the other side. She tapped on the door but nothing happened. So she knocked much harder and was greeted by a loud shout of "Welcome to . . ."

The Violent Interstellar Medium

Alice was amazed at what she saw. The room was a complete mess. It was very dusty and full of smoke. In some places there were huge piles of dust, which were clearly very cold. Electric radiators were scattered here and there and a few infrared and ultraviolet lamps shone at all sorts of curious angles. Every few minutes someone set off a firecracker. The animals in the room ran around from chair to chair and bumped into each other with terrific crashes. Perhaps worst of all was the terrible smell which permeated everything. The odor reminded Alice of a combination of all the most horrid gases which could be produced in a practical chemistry laboratory.

Alice choked and said to a passing animal, "Whatever is going on here?"

"Ask the boss," the animal said. "He'll appear in a moment."

And sure enough, materializing out of nothing came none other than the Cheshire Cat[1]. As usual, his grin appeared first and then the rest of him. Throughout their discussion, in fact, he kept appearing and disappearing and he seemed very much at home in this incredible muddle.

"It's so nice to see you again, Cheshire Cat," said Alice. "Could you explain to me exactly what is going on in here?"

"Gee, Welcome, Sweety," said the Cheshire Cat. "What's a nice girl like you doing in our interstellar workshop? Here we simulate the ecosystem of the interstellar environment—the space between the stars. By living under interstellar conditions, we aim to sensitize the students' perception of the modularities of biological and nonbiological evolution. The students use correlated developmental communication to perform an emotive diagnostic evaluation of advanced democratic objectivity. They use ultra procedural techniques to parameterize the on-going unstructured resources of the interstellar components. Cognitive . . ."

READY

RUN

correlated developmental communication
emotive diagnostic evaluation
advanced democratic objectivity
creative democratic disfunction
emotive implicit rationalisation
correlated empirical re-organisation
multi-socioeconomic rationalisation
ultra procedural techniques
on-going unstructured resources
emotive diagnostic components
extra-psycholinguistic autonomy
programmed meaningful maladjustment

READY

THE JARGON PHASE GENERATOR
(from "The Computer Book," BBC Publications)

```
  5 REM JARGON PHRASE GENERATOR
 10 REM RESERVE SPACE FOR ARRAYS
 20 DIM a$(24)
 30 DIM b$(24)
 40 DIM c$(24)
 50 DATA basic,divergent,programmed,operational,affective
 51 DATA child-centred,multi-,emotive,disadvantaged,on-going
 52 DATA informal,ultra,interdisciplinary,cognitive,relevant
 53 DATA correlated,extra-,innovatory,viable,supportive,elitist
 54 DATA micro-,creative,advanced
 60 REM FILL FIRST ARRAY WITH DATA
 70 FOR x = 1 TO 24
 80 READ a$(x)
 90 NEXT x
100 DATA meaningful,procedural,significant,democratic
101 DATA sociometric,consultative,empirical,unstructured
102 DATA implicit,perceptual,psycholinguistic,coeducational
103 DATA reactionary,motivational,academic,conceptual
104 DATA socioeconomic,hypothetical,ideological,theoretical
105 DATA developmental,compensatory,diagnostic,experimental

110 REM FILL SECOND ARRAY WITH DATA
102 FOR y = 1 TO 24
130 READ b$(y)
140 NEXT y
150 DATA situation,over-involvement,evaluation,components
151 DATA disfunction,methodology,quotients,re-organisation
152 DATA rationalisation,activities,communication,resources
153 DATA synthesis,validation,techniques,consensus
154 DATA maladjustment,sector,criteria,autonomy,analysis
155 DATA polarisation,objectivity,strategy
160 REM FILL THIRD ARRAY WITH DATA
170 FOR z = 1 TO 24
180 READ c$(z)
190 NEXT z
195 REM GENERATE 12 JARGON PHRASES
200 FOR m = 1 TO 12
205 REM CHOOSE 3 RANDOM NUMBERS
210 LET a = RND(24)
220 LET b = RND(24)
230 LET c = RND(24)
235 REM PRINT RANDOMLY SELECTED WORDS FROM
    ARRAYS
240 PRINT a$(a);"   ";b$(b);"   ";c$(c)
250 NEXT m
360 END
```

"Stop!" shouted Alice. "You are using the Jargon Phrase Generator from the BBC's Computer Book[2]—I've seen that program, too."

Anyway, it was now clear to Alice what all the mess was about. She remembered all those lectures about the violent interstellar medium [3]. Dust particles, similar in size to the particles of cigarette smoke, and gas are

The Horsehead Nebula lies in the constellation of Orion. It is
part of a huge region of active star formation, and this picture
shows some of the complexity of these regions. It is within the
darkest regions that the youngest stars and protostars are found.

The Orion Nebula is optically the most conspicuous part of the
star formation region in the constellation of Orion. The clouds of
hot gas seen in this picture are excited by four young blue stars,
known as the Trapezium stars, which lie within the brightest part
of the nebula.

Carbon Monoxide   Methanol   Ethanol   Formaldehyde   Acetaldehyde   Dimethyl Ether   Formic Acid   Methyl Formate   Acetylene   Propyne   Thioformaldehyde   Methyl Mercaptan   Hydrogen Cyanide   Carbon Monosulphide   Carbonyl Sulphide   Formamide   Ketene   Methyl Cyanide   Vinyl Cyanide   Ethyl Cyanide   Methylamine   Cyanamide   Methyl Cyanoacetylene   Hydrogen Isocyanide   Cyanoacetylene   Cyanobutadiyne   Cyanohexatriyne   Cyanooc-tatetrayne   Cyanodecapentayne   Methylenimine   Isocyanic Acid   Isothiocyanic Acid   Diatomic Carbon   Hydrogen   Ammonia   Water   Hydrogen Suphide   Nitric Oxide   Nitric Sulphide   Sulphur Monoxide   Sulphur Dioxide   Silicon Monoxide   Silicon Monosulphide   Nitroxyl Hydride   HydroxylMethylidene   Cyanogen   Formyl   Ethynyl   Butadiynyl   Cyanoethynyl   Methylidene   Formylium   Thioformylium   Diazenylium

Alice thinks about the long list of chemicals now known to be present in the interstellar medium.

present throughout the space between the stars. This medium is in constant turmoil, being buffeted by stellar explosions and the winds from stars of all sorts. In cold dusty regions, new stars are formed. Nothing really stays in one place for very long at all. The medium is constantly changing and there are all sorts of sources of heat and many regions where the gas cools throughout the volume. All sorts of gases are present in the interstellar medium, many of them consisting of molecular gases like formaldehyde, hydrogen sulphide, carbon monoxide, ammonia, and lots and lots of others. This explained the terrible smells which permeated the Cheshire Cat's laboratory. Alice also remembered that ethyl alcohol, or ethanol, as she had been taught to call it in her organic chemistry classes, was present in large quantities—which may have had something to do with the somewhat disorderly behavior of some of the animals in the room. The fact that nothing lasted very long in one place made it the ideal place for the Cheshire Cat, who kept appearing and disappearing.

The Vela supernova remnant. The red filaments are the remnant
of a star which exploded about 10,000 years ago. The whole star
was disrupted and ejected into the interstellar medium at a high
velocity. Close to the center of the nebula is a pulsar, which
flashes at a rate of about 12 times per second. The pulsar is a
magnetized rotating neutron star which formed during the
collapse and explosion of the presupernova star.

He reappeared beside Alice and said, "Our aim is to develop a spaced-out life-style whereby the processes of procreation, birth, and death of stars can be simulated under laboratory conditions."

Through the dust Alice could see the animals stumbling around in a random manner, and there were certainly some very strange goings-on in the star formation corner. She was rather shocked.

The Cheshire Cat beamed. "They are all a bit excited, but so would you be if you had to get into the spirit of feeling like a bit of interstellar gas!"

"Well," said Alice in a firm tone. "I really cannot approve of the behavior of some of your students. If they want to display their affection for each other, they could at least wait until after office hours!"

"O.K., kids! Cool it!" shouted the Cheshire Cat.

"Tell me, Cheshire Cat," asked Alice. "Do you understand how stars are formed?"

"Not really," he answered. "There are just so many basic things we don't really know or understand[4]. For example, we don't know what it is that determines the rate at which stars form out of the interstellar gas or how this rate depends upon the density, temperature, and chemical composition of the gas. We don't really know how many massive and how many light stars are created and how that depends upon the properties of the gas. We would like to know the precise sequence of events which takes place when a new star forms, but we don't. It's all very annoying."

Alice could only agree. Understanding how stars are formed is perhaps the most important unsolved problem of the astrophysics of stars and galaxies. To have a proper understanding of the evolution of galaxies as a whole, one just has to know how stars are born.

Many of the most exciting observations of the very youngest stars have been made in the infrared waveband, where dust becomes transparent and the very youngest stars and perhaps even their precursors, the protostars, emit very intense radiation. Infrared observations will not be possible with the first generation of scientific instruments on board the Space Telescope, but observations with the telescope will undoubtedly throw new light on many of the basic problems which the Cheshire Cat had described. One of the most powerful ways of doing this will be by observing stellar systems of widely differing ages.

There was a pause and then Alice saw some of the animals stretching huge rubber bands throughout the room. They tangled them round chairs and tables and lights and threaded them through the piles of dust in various nooks and crannies of the room.

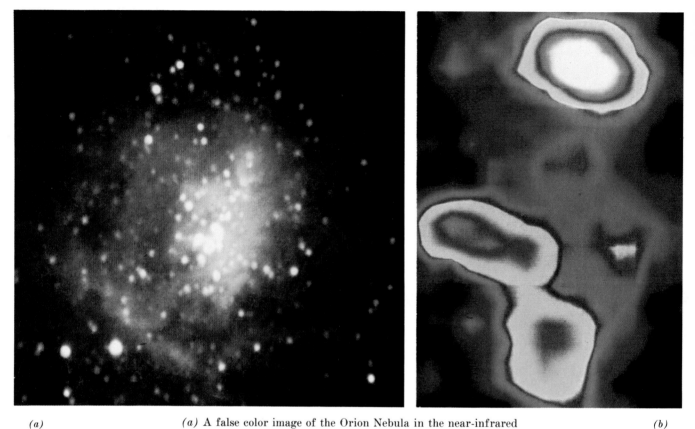

(a)

(b)

(a) A false color image of the Orion Nebula in the near-infrared waveband 1 to 2.2μm. Hot unobscured objects are shown as white, whereas cooler or obscured objects are shown as yellow. The four bright white stars in the center of the picture are the Trapezium stars, the source of energy for the nebulosity seen on page 38. The yellow stars to the north of this region are much cooler objects, which lie behind the Orion Nebula. These stars are much younger than the Trapezium stars and some of them may be protostars—objects in whose cores nuclear burning has not yet begun. *(b)* A false color image of the region of the yellow stars seen in *(a)* but now as observed at the far-infrared wavelength of 10μm. The intense source to the north is known as the Becklin-Neugebauer object and is an extremely powerful source of far-infrared emission. New stars may be forming inside some of these sources.

"This," explained the Cheshire Cat, "completes our simulation of the interstellar medium. These rubber bands represent the magnetic field which is present in the interstellar gas[5]. All we have to do now is to stick everything in the room to the rubber bands with syrup and then we will have the perfect analogue for the interstellar medium. Right, kids, let's go!"

A sketch of the violent interstellar medium. The dark regions are cold, dusty clouds inside which stars are forming. The medium as a whole is buffeted by the explosions of stars. These heat up large regions of the interstellar gas and the ejected shells of material may collide with one another. In this way, extensive regions of hot gas forming "tubes" through the interstellar medium may well be formed.

Alice rushed out of the room. It was bad enough being exposed to smoke, dust, and noxious chemicals, but she would not be tied up in syrup-coated rubber bands. Once she was outside, however, she had to admit that this was a rather good model for the medium between the stars. The medium is very chaotic, and a magnetic field ties all the components together just like these horrid syrupy rubber bands. The Space Telescope will do wonders for solving so many of the problems which this picture provokes. It will show where the hot gas is located; the chemical composition of the gas in a wide variety of different regions can be worked out so that astronomers may be able to understand where all the elements came from in the first place. The process of circulating the interstellar gas through stars will be studied particularly well with Space Telescope. With the cameras, it might even be possible to watch the precursors of stars, the protostars, collapsing or to watch the very youngest stars blowing off the dusty gas clouds in which they are formed. She hoped the Cheshire Cat and his students would recover in time to make these exciting observations with the Space Telescope.

The Building Blocks of the Universe

After all that chaos, Alice felt like something rather less violent. She had just decided to look for some other scientists who might help her with her essays when she saw a sign on another door which read

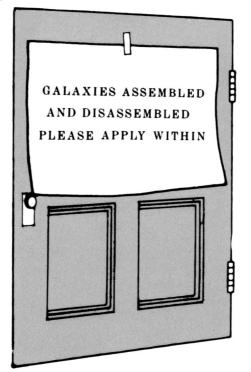

GALAXIES ASSEMBLED AND DISASSEMBLED PLEASE APPLY WITHIN

Alice realized that she was going to larger and larger scales in the Universe[1]. The galaxies are the building blocks of the Universe[2]. Our own Galaxy is but one of millions, each one made up of millions or billions of stars. They are among the most beautiful images in astronomy. Galaxies come in all sorts of shapes and sizes. There are the giant spiral galaxies which must be rather similar to our own; the giant elliptical galaxies, some of which are

The giant spiral galaxy nearest to our own, the Andromeda
Nebula, or the Andromeda Galaxy, is probably not too different
from our Galaxy.

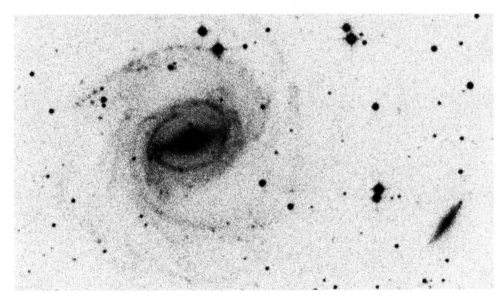

The barred spiral galaxy NGC 3313.

The giant elliptical galaxy M87, which lies close to the center of
the Virgo cluster of galaxies.

The nearby giant elliptical galaxy NGC 5128. This galaxy is a
strong source of radio emission. The main body of the galaxy is
crossed by a prominent dust lane.

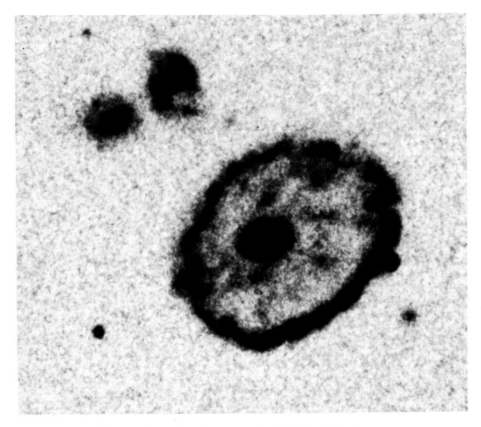

The peculiar galaxy known as the Cartwheel. Its strange
appearance is almost certainly due to a recent collision or to
interaction with one of its nearby companions.

very powerful sources of high energy particles; and the peculiar galaxies in
which some dreadful accident seems to have occurred. Alice could understand why they were such exciting things to study but she was also well aware
of the fact that they pose some of the most difficult problems of modern
astronomy.

This was just too good an opportunity to miss. Alice remembered vividly
two striking things about the study of galaxies with the Space Telescope.
First, she remembered the very dramatic improvement in angular resolving
power which will be obtained for the study of all galaxies. The extra factor of
ten in definition will enable all sorts of studies to be made of galaxies which
are ten times further away than those which can be studied from the ground.

Second, Alice remembered that the cameras on the Space Telescope will
enable astronomers to look so far into space that they will typically observe

(a)

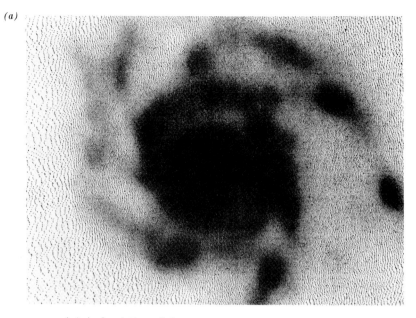

(a) A simulation of the appearance of a galaxy as observed under good observing conditions from the ground. *(b)* The same galaxy, but as it will be observed by the Hubble Space Telescope at the same wavelength. Notice that not only is much more detail of the galaxy visible but very faint stars, which are not apparent in (a), can also be seen. These images make the important point that the key advantages of the Space Telescope are its very high angular resolving power and its great ability to detect faint starlike objects.

(b)

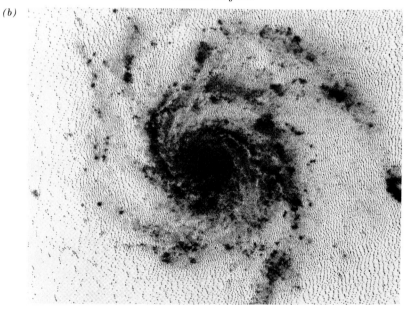

galaxies which emitted their light when the Universe was about half its present age. The words of her professor came back to her: "Every deep-sky picture taken with the Space Telescope will be an image of the Universe in the distant past and not a picture of the Universe as it is now"[3].

Alice peered round the door and there were some of her old friends talking about galaxies. The Mad Hatter, the March Hare, and the Dormouse were taking tea with bread and butter much as before[4]. Alice did not wait for any formal introductions but said, "I am afraid I'm in a bit of a hurry with my essays so could you please tell me what you think are the most important things which the Space Telescope will achieve in the study of galaxies?"

The Mad Hatter said excitedly, "Excellent! Let's play riddles! Well, you begin and then we ask you."

Alice thought for a moment and then said, "What parameters define the properties of a galaxy besides its mass?" She remembered that this was a rather interesting idea which may simplify many problems of the astrophysics of galaxies[5]. It may be that what distinguishes galaxies of different types is simply the environment in which they are located. It is an intriguing idea which needs much more study. It certainly seemed a good opening question for the riddles.

"That's not very funny," said the March Hare. "Ask us another!"

"What are the dynamical properties of the stars in the elliptical galaxies?" said Alice. She remembered the remarkably beautiful work which had been done to show that the stars in galaxies do not behave at all like the particles in an ordinary gas[6]; they behave more or less independently of one another and are only influenced by the gravitational field of the galaxy as a whole.

The motions of the stars may contain information about where they first came from and how the galaxy was first formed.

"Good," shouted the March Hare. "Now give us your third question."

"What is the nature of the dark matter in galaxies?" said Alice hopefully. This was one of the biggest puzzles of modern astronomy[7]. Large systems in the Universe, like giant galaxies and groups and clusters of galaxies,

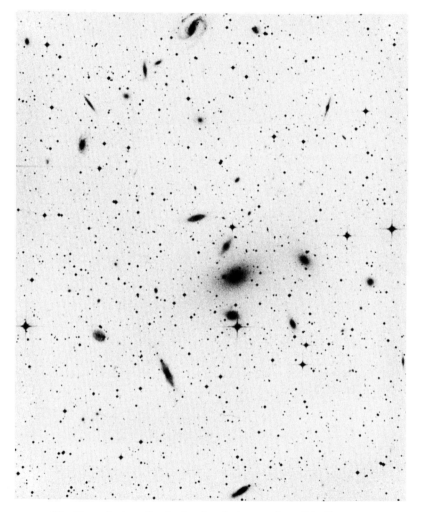

The Pavo cluster of galaxies. In clusters such as this there must be about 10 times as much mass present as can be inferred from the visible parts of the galaxies present. In the center of this cluster is a dominant giant galaxy, which may have grown by consuming other members of the cluster.

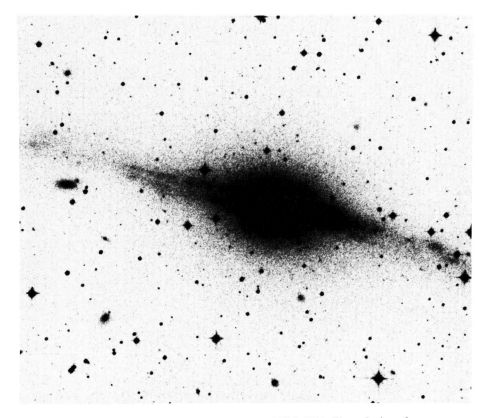

The peculiar lenticular, or SO, galaxy NGC 5084. The velocity of
rotation of this galaxy in its outer regions is very large, which
indicates that it must contain about 10 times the mass that would
be expected of a normal spiral galaxy.

contain much more mass than astronomers would infer from their visible
light. Clusters of galaxies are particularly worrying because the discrepancy
seems to be about a factor of ten for some systems which had been studied
intensively.

"Congratulations! Well done!" said the Mad Hatter. "An excellent set of
riddles—we give you first prize."

"But," said Alice, "Aren't you supposed to answer the riddles?"

"Not at all," said the Mad Hatter. "In Wonderland, you get prizes for
asking the questions, not for giving the answers. It's just like extragalactic
astronomy."

Alice had met this sort of answer before and was not going to be taken in by
it. "I don't think you know the answers anyway. I think you are just bluffing.
Do try to give me an answer to at least one of them."

The Mad Hatter and the March Hare were taken somewhat aback and nudged the Dormouse, who woke up and took a sheet of paper out of the teapot. He coughed and then read sleepily from the sheet of paper.

"Here is a list of the various types of hidden mass which could be present in clusters of galaxies to make up the hidden or dark mass," said the Dormouse.

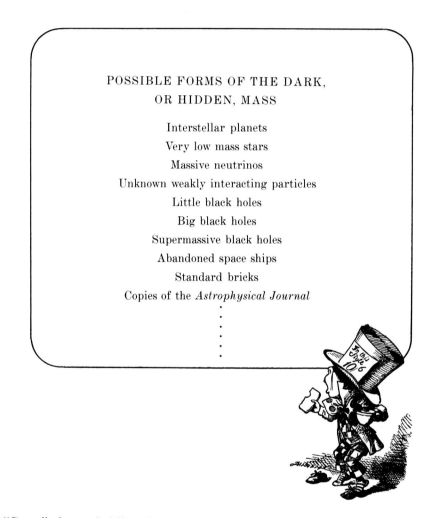

POSSIBLE FORMS OF THE DARK,
OR HIDDEN, MASS

Interstellar planets
Very low mass stars
Massive neutrinos
Unknown weakly interacting particles
Little black holes
Big black holes
Supermassive black holes
Abandoned space ships
Standard bricks
Copies of the *Astrophysical Journal*
.
.
.
.
.

"Stop," shouted Alice. She recognized that the Dormouse was being serious; indeed, there is a huge range of possible types of dark matter in the Universe. She remembered a key fact about a purely observational science like astronomy. There are practical limits to what can be observed in any waveband and certain forms of matter are very difficult to detect. Basically, if objects don't emit much radiation, they can be very difficult to observe.

"I think I know the answers I need," thought Alice. "Space Telescope will enable us to look for some of the types of matter which could contribute to the hidden mass in galaxies and clusters of galaxies—things like low-mass stars, planets, and others forms of coldish matter which are difficult to detect. In addition we should be able to find out how severe the hidden mass problem is in a wide range of different systems using the spectrographs on the Space Telescope."

She was about to take her leave when they all shouted "Wait! We have to ask you questions now—that's the rule of riddle playing."

"Oh dear," thought Alice. "I knew I couldn't get away without this."

"What is the most common answer in extragalactic astronomy?" asked the March Hare. Alice thought hard but was unable to think of anything. "I don't know," she said.

"Quite correct!" said the March Hare.

"What is the difference between the Andromeda Nebula and the Cheshire Cat?" asked the Mad Hatter.

Alice said, "I don't know. They seem pretty far out to me."

"Correct again!" said the Mad Hatter.

"Why is a big fish like the biggest galaxy in a cluster of galaxies?" asked the Dormouse.

Alice was delighted—she knew the answer. "They both eat smaller objects and grow fat on them!"

"Excellent!" said the March Hare. "Your prize is that we will sing you our song about galactic cannibalism. Do you know the song called "The Hausman and the Ostriker"?[8] We learned it from Tweedledum and Tweedledee, who live next door. You must see them soon."

Alice gritted her teeth and said politely, "I would love to hear your song."

They began at once.

THE HAUSMAN AND THE OSTRIKER

The sun was shining in the sky
    The snow was on the ground,
Welcome to sunny Baltimore
    And institutes around—
And this is very strange because
    They were in Princeton town.

The Hausman and the Ostriker,
  They sat in Peyton Hall,
The Hausman was a student
  The Ostriker ruled it all—
"We really must get working
  On your project ere the fall!"

"I love all astrophysics,"
  The Ostriker truly said.
"Gas and galaxies, stars and dust,
  Optical, infrared"—
He drank a cup of coffee
  And ate a slice of bread.

"I love the little galaxies,
  They look so very nice.
I'd like to know how they evolve
  And solve it in a trice":
"You'll be lucky" the Ostriker said.
  "Pass me another slice."

"I have it!" said the Ostriker,
  "This lunch gives me the clue.
Galaxies must eat galaxies,
  It really must be true—
Computerize the problem!
  That's just the job for you!"

Computer code was written
  (They both are very bright).
They put in many galaxies
  Some massive and some light.
"Now let them go and see
  Who is winner of the fight."

The galaxies rushed together
  To make a cluster large and fine.
They zoomed right through the center
  In a shortish space of time:
"That's violent relaxation,
  What a horrible word to rhyme!"

But see the very largest one,
   He spirals to the middle,
Flinging off the little chaps
   And causing them to giggle.
"We must come here again," they said.
   "And have another jiggle!"

So in they came in orbits bound
   Towards the largest fellow.
"Oh dear!" they cried. "He's grown fat
   I feel like fresh-made Jello—
Dynamical friction slows me down,
   I feel distinctly yellow!"

And so it was the greedy giant
   Ate galaxies large and small
He grew into a supergiant
   And dominates them all—
The story ends about this point
   The verse begins to pall.

The Hausman and the Ostriker,
   They wept to see them die.
"It is so sad it had to be
   But it happens every try":
"Don't worry," said the Ostriker.
   "Now try some apple pie."

Alice had to agree that this is a very beautiful explanation of the observation that in giant clusters of galaxies, there are very often supergiant elliptical galaxies at their centers and that this is almost certainly what is observed in some nearby clusters. Jim Gunn had even named one picture "a cD galaxy at lunch"[9].

Interactions or collisions between galaxies has become a subject of the greatest interest for many aspects of astronomy[10]. The *Infrared Astronomy Satellite, IRAS*, had discovered that many of the galaxies which are the strongest emitters in the far infrared waveband are interacting galaxies in which one galaxy has collided with another. The very high angular resolution

A rich cluster of galaxies, discovered by Dr. James Gunn and his colleagues, in which the central dominant galaxy consists of a number of separate condensations. It is likely that this system is in the process of forming a giant galaxy at the center of the cluster.

capability of the Space Telescope will enable astronomers to investigate the processes of cannibalism and interactions between galaxies at very great distances. It is also likely that interactions between galaxies are important in providing the fuel for active galactic nuclei, and images taken with the Space Telescope will show directly whether or not these ideas are correct. For example, quasars are almost certainly the nuclei of very distant and active galaxies. One might expect to see the debris of the galaxy which had interacted with the quasar's host galaxy; the Space Telescope should be able to see these effects in detail.

"Thank you very much!" Alice said. "I really must be rushing."

The last thing she heard was a cry of "Watch out for the black holes!"

Active Galactic Nuclei and Black Holes

Alice could not help but feel a little apprehensive about the prospect of bumping into a black hole and so she walked very warily along the corridor to the room marked

BLACK HOLES—WARNING—
THE SURGEON-GENERAL
HAS DETERMINED THAT
BLACK HOLES ARE
DANGEROUS TO YOUR
MIND AND BODY

In the middle of the room[1] sat the Caterpillar[2] on a large mushroom smoking his pipe and looking very solemn. Alice remembered that he was a rather fierce animal who asked all sorts of penetrating questions.

The Caterpillar took a long puff at his pipe. "Young person!" he said. "Why do you think black holes have anything at all to do with active galaxies?"

"Well," said Alice, "the most important thing we observe about the centers of some active galaxies and quasars is that they emit huge amounts of energy in a very short time[3]. For example, in the most active galactic nuclei, the nucleus must liberate more than a billion times the luminosity of the Sun in a few days. This means that there must be a very small object producing huge amounts of energy very rapidly. The simplest way of explaining this is to suppose that there is a very massive black hole[4] in the middle of the galaxy and that matter falls into it. As the matter is most unlikely to fall straight in, it has to lose its rotational energy, and it is this process which produces the large bursts of radiation we see from the nuclei of active galaxies and quasars."

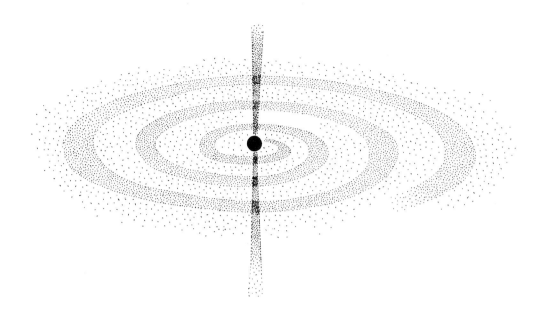

A schematic picture of the very nuclear regions of an active galaxy. At the center itself is a black hole and about it is a disk of material which slowly drifts in towards the center. This infall of material into the nucleus is known as accretion and is a very powerful energy source. Besides producing the huge releases of energy observed in the optical, ultraviolet, and X-ray wavebands, this process may well be responsible for the beams of particles which are observed to be ejected from some active nuclei.

"Have you actually seen a black hole?" asked the Caterpillar.

"Not really," said Alice. "You cannot actually see the black hole because nothing that falls in, including light, can ever come out again. That's why it's called a black hole."

"Well, if you cannot ever see black holes, how can you ever know if they exist?" retorted the Caterpillar.

"According to my teachers, there is nothing else that could be reasonably proposed to explain the power of active nuclei. Other sources of energy are much less efficient and seem very contrived."

"And you think building models of objects which you will never be able to see is not contrived. I call that cheating!"

Alice was beginning to become a bit upset. "Well, why are you studying these objects if you don't believe in them?"

"Who said I had to believe in something to study it?" said the Caterpillar. "I make it a point of principle to consider the most outlandish theories because, although they are perhaps unlikely to be generally true, they must be applicable *somewhere* in the Universe. As the White Queen told you long ago, some of us can believe as many as six impossible things before breakfast"[5].

Alice had been warned about some of the mathematicians and astronomers who worked on black holes. She could, however, appreciate the point of his argument. The assumption that there are massive black holes in the nuclei of active galaxies is not so different from the sort of assumptions one often has to make in astronomy. Astronomers and physicists love elegant and economical theories, and it is remarkable how often they are indeed correct. There is no fundamental reason, however, why the theories which explain astronomical phenomena like active nuclei must be elegant or simple.

In fact, Alice was certain in her own mind that black holes do exist. We know that neutron stars exist because of the existence of pulsars and X-ray binaries. If a neutron star were to be squashed by a factor of only three in radius, it would become a black hole. It seemed perfectly natural that some stars should form black holes rather than neutron stars at the end of their lifetimes, when all other sources of internal support had been exhausted.

"Well, all I can say is that I find the story of the galaxy NGC 4151[6] very exciting and convincing," said Alice. "It seems to me to be exactly the sort of problem which is ideally matched to the capabilities of the Space Telescope."

"Enlighten me, child," said the Caterpillar.

"NGC 4151 is the nearest example we know of a hyperactive nucleus. In a long exposure photograph, it looks just like an ordinary spiral galaxy, but if you take only a short exposure, you find that there is a brilliant starlike object in the nucleus itself. Some European astronomers used the *International Ultraviolet Explorer*[7] to study this object and discovered that there are rather rapid time variations in the optical and ultraviolet radiation from the nucleus. This intense radiation excites the surrounding gas clouds and causes them to emit strong line radiation at the resonance frequencies of the atoms and ions in the clouds. Since the astronomers were able to measure the time delay between the lighting up of the nucleus and the excitation of the gas clouds, they could work out how far away from the nucleus the gas clouds had to be. They also knew the velocities of the clouds from the breadth of the emission lines. They could then use Kepler's laws of planetary motion[8] to work out the mass of the nucleus itself. This turned out to be about one billion solar masses, exactly what theoreticians say we need to explain the huge amounts of energy emitted from the nucleus."

"I suppose you want the Space Telescope to observe lots of other active galaxies," said the Caterpillar.

(a)

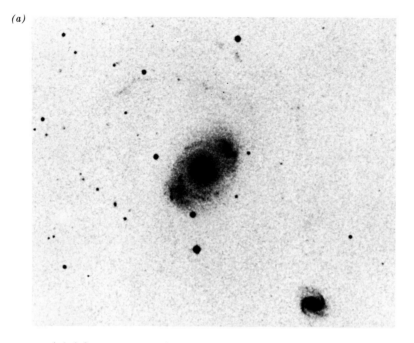

(a) A long exposure photograph of the galaxy NGC 4151, which appears to be a spiral galaxy. *(b)* A short exposure photograph of NGC 4151. Its nucleus appears to be stellar. The fact that the nucleus must be very compact is confirmed by the observation that its brightness varies.

(b)

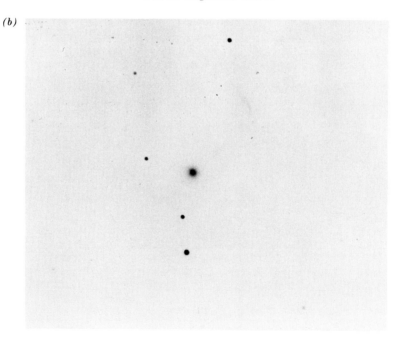

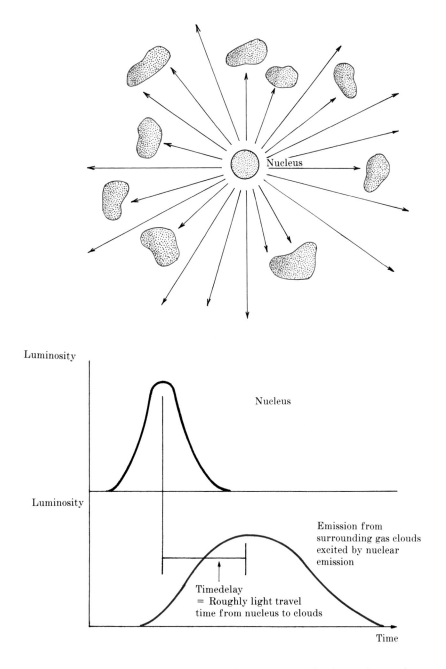

A model of an active galactic nucleus showing how a burst of radiation in the nucleus can lead to subsequent excitation of the surrounding gas clouds, which will be observed to glow at a later time.

"Oh, yes! It's a fantastic opportunity. Only the Space Telescope will be able to study systematically the whole range of active nuclei—the Seyfert galaxies, the radio galaxies, the quasars, the BL-Lacertae objects[9]—all the most exotic objects in astronomy will be studied as never before. We may actually begin to have real tools with which to study the regions closest to the black hole itself."

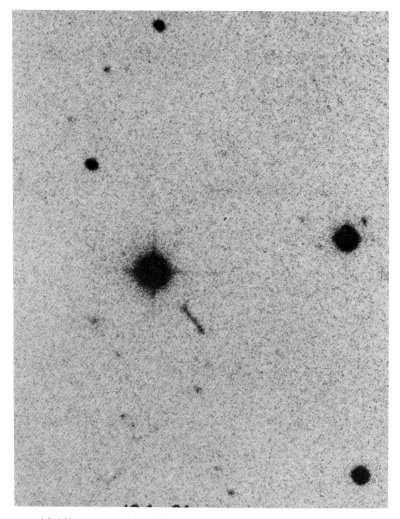

3C 273 was one of the first quasars to be identified and is still the brightest known object of this type. Although it looks like a star in this picture, it is in fact very distant. Some faint galaxies can be seen close to the quasar; these are at the same distance as the quasar. One remarkable feature of 3C 273 is the jet which extends from the quasar towards the bottom right of the picture.

"I really don't know why you think you need our help with your essays. You seem to know all about active nuclei already," said the Caterpillar.

"What are you studying here?" asked Alice. She suddenly noticed that the room was very smoky and that all the animals were smoking just like the Caterpillar. They seemed to be spending most of the time leaning back in their chairs and blowing jets and smoke rings.

"Oh, lots of things. Besides those you've mentioned, we are trying to understand how black holes generate high-energy particles and magnetic fields. One of the great discoveries of modern astronomy has been that somehow the nuclei of active galaxies are able to generate very high energy particles with energies similar to and much greater than the most energetic particles which we can produce in accelerators on Earth[10]. Nobody really understands how they are able to do this.

"We are also trying to understand how the black hole ejects these particles in beams from the center to produce the lovely radio maps produced by radio telescopes like the Very Large Array. My colleagues here are producing jets and trying to understand how they interact with the surrounding medium. Some of them can produce the most beautiful smoke rings. Every time I smoke my pipe, I am performing an experiment in the magneto-gas-dynamics of beams of ultrarelativistic gas produced in active nuclei."

Alice thought to herself that this was the first time she had heard any positive reason for smoking. She had to agree, however, that these jets were

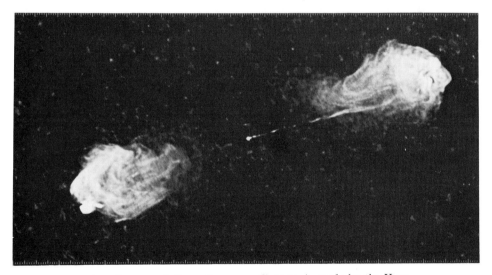

A radio map of the radio source Cygnus A, made by the Very
Large Array.

among the most important phenomena to be studied by the Space Telescope. It may well be that there are jets associated with most active galaxies. It will be very important to know if they possess beautiful jets like the well-known one in the center of the giant elliptical galaxy M87[11]. The radio astronomers had discovered enormous versions of these jets and also very tiny versions of the same phenomenon close to the nuclei themselves. Some of these seemed to be moving outwards from the nucleus at speeds greater than the velocity of light—this seemed to be an ideal topic for the Caterpillar and

The central regions of the giant elliptical galaxy M87 showing the well-known jet emanating from the nucleus.

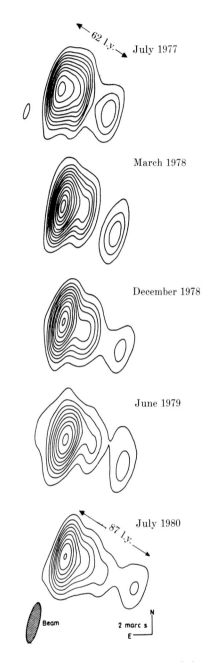

A sequence of images of the radio structure of the central
regions of 3C 273 taken at different epochs and made by the
technique of Very Long Baseline Interferometry (VLBI). It can
be seen that in the space of 3 years the outer component
separates from the center by 25 light-years, indicating that the
apparent transverse velocity of the component must be about
eight times the velocity of light.

his students[12]. These jets had, however, proved very much more difficult to find by optical means. The Space Telescope will look deep into galaxies and will have a unique capability for studying these jets and possibly even for measuring directly the speed at which they move out from the nucleus. This would be a terrific achievement. To be honest, no one really knows what role these jets play in the evolution of galaxies, but they are probably associated in some way with the processes of energy and angular momentum transport in the innermost regions of galaxies.

"Well, thank you," said Alice.

"Just keep away from the black hole garbage bin by the door as you leave," said the Caterpillar. "It's very useful for getting rid of theoretical papers and weak students!"

Alice thought it best to escape quickly. She ran out into the corridor and found that there was only one room left. On it was written in large letters the single word

Big Bangs and Little Bangs

Alice took a deep breath. So far she had been talking with conventional astronomers, and look at the problems she had had getting sense out of them! Whatever would it be like when she wanted to find out about cosmology, which is a notorious subject for controversy and hot dispute![1] She paused for a moment to clear her head and tried to decide how to tackle the problem. One thing was certain—she was not going to be diverted into nonconventional astronomy. She would concentrate on the Hot Big Bang model of the Universe[2]. She remembered that there are three very good reasons why the Hot Big Bang provides such a satisfactory framework for cosmological discussions[3].

1. The Universe must be expanding uniformly at the present time for two reasons—first, it is observed that the farther away a galaxy is from us, the greater is its velocity recession away from us (Hubble's Law) and, second, because the Universe looks the same in all directions on a large scale.

2. The microwave background radiation permeates all space and has an almost perfect thermal spectrum. This means that all the matter and radiation in the Universe were once in thermal equilibrium.

3. The observed abundances of the light elements such as helium, deuterium, and lithium can be accounted for naturally by element formation in the early stages of the Hot Big Bang, when the Universe was expanding very rapidly. These light elements cannot be explained by element formation in stars.

The great attraction of the Hot Big Bang model is that these three independent large-scale features of the Universe can be explained by a single theory. As such it provides the most convincing framework within which to study cosmological problems. For Alice, cosmology and the Space Telescope

COSMOLOGY

1

What is the present rate of expansion of the Universe?

2

What is the age of the Universe,
and what is its present rate of deceleration?

3

Is the present deceleration of the Universe
due entirely to the gravitational influence of matter
and radiation in the Universe?

4

Why did radio sources and quasars
show such rapid changes in their properties when
the Universe was younger?

5

Did galaxies form early or late in the Universe?

6

How did galaxies first form?

7

What is the chemical abundance of the material out of
which galaxies first formed?

8

How did the Universe begin?

therefore meant the study of the properties of the Hot Big Bang and in particular of the great problems concerning how everything in the Universe was formed. She made a list of the things she would have to consider. It turned out to be rather long.

"Professor Longair!" she said. "This just shows where your interests really lie—you have biased the whole story towards cosmology! I think that is unfair."

"Alice, you are not allowed to criticize the author—*I* decide what you do and what questions you ask."

"Why don't you answer the questions yourself then?" said Alice. "*You* go into that room and deal with these crazy cosmologists."

"Alice, I don't want to exert any undue influence, but you must realize that you are not in a very strong position; I could change some parts of the story. Let's look on the brighter side. This is the last room and after that I promise you will meet an old friend. You are almost finished."

In she went and there were her two old friends Tweedledum and Tweedledee[4]. "Right," said Alice. "I want brisk answers to these questions and I warn you, I want no nonsense!

"First question. What is the present rate of expansion of the Universe? In other words, what is the value of Hubble's constant, which appears in the velocity-distance relation for galaxies?" [5]

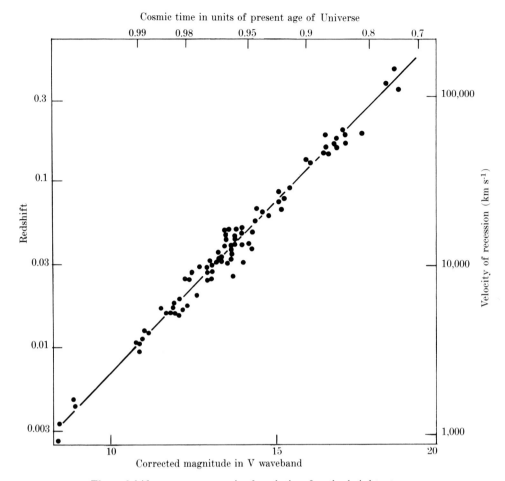

The redshift-apparent magnitude relation for the brightest
galaxies in rich clusters of galaxies. The straight line is the
expected relation if the Universe is expanding uniformly and if
all the galaxies have the same intrinsic luminosity.

"Hubble's constant is 50 kilometers per second per megaparsec," said
Tweedledum.

"Contrariwise, Hubble's constant is 100 kilometers per second per megaparsec," said Tweedledee.

"He's wrong," said Tweedledum.

"He's wrong," said Tweedledee.

This seemed to be business as usual. Alice knew exactly how to handle this
situation. "Tell me," she said, "How accurate do you think your values are?"

Tweedledum and Tweedledee glanced rather nervously at each other.
"Better than 25 percent," they both replied optimistically.

"Well," said Alice, "I believe there is no significant difference between

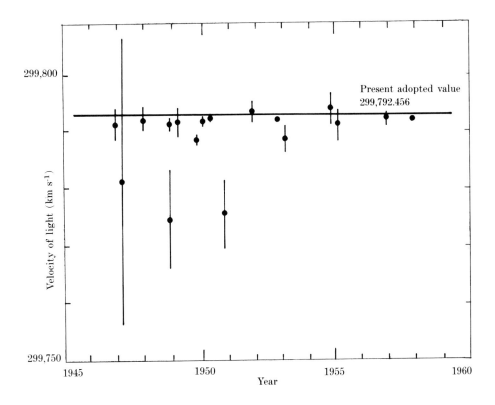

Precise measurements of the velocity of light published in the
period 1945 to 1960. Notice how often the values disagree both
outside the formal uncertainties quoted by the experimenters and
also with the present adopted value.

these values. Let me demonstrate—do you agree that physicists ought to
know how to calculate their experimental errors?"

There was a long pause and they replied cautiously, "Maybe."

"Well," said Alice, "Let me show you a series of successive measurements
of the velocity of light. These measurements were made between 1945 and
1960, and only the best experiments have been selected for inclusion on the
diagram[6]. You will notice that even in these laboratory experiments, there
are very often disagreements well beyond the quoted errors on each measurement. If physicists cannot get their errors right, is it reasonable to expect
astronomers to do any better when they do not even have control of their
experiments?"

Tweedledum and Tweedledee just glowered at each other.

"Happily," said Alice, "This is one of the most important programs which
will definitely be undertaken by the Space Telescope. One of its prime goals

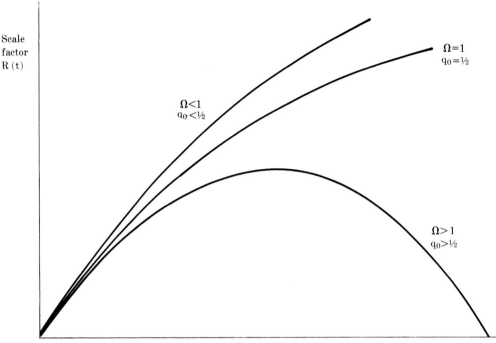

Scale
factor
R (t)

$\Omega < 1$
$q_0 < \frac{1}{2}$

$\Omega = 1$
$q_0 = \frac{1}{2}$

$\Omega > 1$
$q_0 > \frac{1}{2}$

Cosmic time t

The dynamics of different world models. Time runs along the
horizontal axis and the relative size of the Universe (or scale
factor) is shown on the vertical axis. According to the classical
models, the dynamical behavior of the Universe depends only
upon the present density of matter and radiation in the Universe
as described by the density parameter.

will be to measure Hubble's constant and the size of the Universe with
significantly improved precision. There are several ways in which this can be
done and one of them will certainly be by measuring the distances of Cepheid
variable stars[7] in the Virgo cluster of galaxies. Indeed, this was one of the
main astrophysical considerations which led to the decision that the Space
Telescope should not be any smaller than 2.4 meters in diameter.

"Next question! What is the age of the Universe and what is its present rate
of deceleration?[8] You may simply tell me your preferred value for the
deceleration parameter q_0, which is simply a measure of the present rate at
which the expansion of the Universe is slowing down. If the value of q_0 is less
than ½, the Universe will expand forever. If, however, the value of q_0 is
greater than ½, the deceleration of the Universe is sufficiently great that it
will eventually stop expanding and then collapse back to a 'Big Crunch.' Of

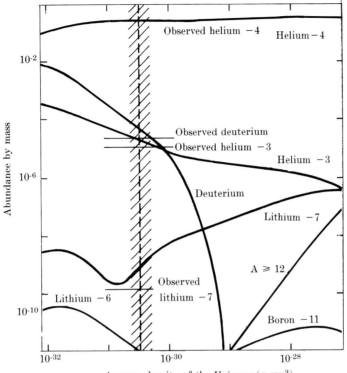

The observed abundances of the light elements compared with
the predictions of different classical world models. For a range
of low-density world models, the Hot Big Bang models can
explain the observed abundances of many of these elements.

course, this assumes that there are no unknown forces acting on matter on a
very large scale in the Universe besides gravity, that is, no invention of new
physics."

Tweedledum said, "I favor a value of q_0 much less than ½. Since my
preferred value of Hubble's constant is 50 kilometers per second per megaparsec, the age of the Universe is 20 billion years and it will expand forever."

Tweedledee said, "Contrariwise, I favor a value of q_0 of ½, and since my
preferred value of Hubble's constant is 100 kilometers per second per megaparsec, the age of the Universe is 7 billion years."

Tweedledum retorted, "My value of the age of the Universe is consistent
with the ages of the oldest stars we know of, which lie in the range 13 to 17
billion years. The Universe cannot be younger than this and so his value is
incorrect. Furthermore, my value is consistent with all known facts about the

Universe, including its average density, which can be inferred from primordial nucleosynthesis of the light elements."

"Nonsense!" replied Tweedledee. "We know that the deceleration of the Universe must be within about a factor of 10 of the critical value $q_0 = \frac{1}{2}$. There is no logical reason why q_0 should take any value between 0 and infinity, and since it is so close to being $\frac{1}{2}$, it must be $\frac{1}{2}$. Furthermore, his value is only a lower limit to the amount of mass in the Universe, whereas mine is comfortably greater than what we see. He also believes he understands the evolution of stars in globular clusters, and I think he is being very optimistic."

Alice tried to find a way of resolving their differences. "I think my previous comment about uncertainties in astronomy and cosmology is even more relevant to this problem. You are describing some of the most difficult measurements in astronomy. I think you should be pleased that you seem to agree within a factor of about 10. It could well have been otherwise.

"I suspect we have also found the answer to the third question as well. 'Is the present deceleration of the Universe due entirely to the gravitational influence of matter and radiation in the Universe?' We are asking whether the observed deceleration of the Universe can be attributed entirely to the decelerating effect of the amount of mass we observe to be present in the Universe [9]. What you have just told me is that the deceleration parameter probably lies somewhere in the range 0.05 to 1. To produce the light elements we need a matter density now at the bottom end of this range, and as Tweedledee rightly remarked, this must be a lower limit to the amount of mass in the Universe. I think we can say that, within about a factor of 10, the present deceleration of the Universe can be attributed to the gravitational influence of known forms of matter. Many cosmologists would argue that if these independent estimates agree as well as this, then the standard scenario is probably correct and there is no need to go searching for new physics. However, one would like to know the answer to this question much more precisely, and there is no question but that Space Telescope will provide us with improved values of Hubble's constant and the deceleration parameter, provided enough effort is devoted to the problem."

Tweedledum gave Tweedledee a dig in the ribs and laughed. "See, I told you she would agree with me. The value of Hubble's constant is about 50 kilometers per second per megaparsec and q_0 is small. This is consistent with the best physics we have available. You got it wrong as usual."

Tweedledee kicked Tweedledum hard on the shins and shouted, "Oh no, I'm not! She actually said we don't know the answer and that my values are correct within the present uncertainties. You are far too keen on finding the final solution to everything and assume that there is no fundamentally new physics for us to learn from astronomical observations."

Tweedledum and Tweedledee rubbed their wounds and glowered at each other.

The rest of the questions were all about the evolution of the Universe[10]. How did objects differ in the past from what they look like now? How did galaxies and the Universe itself come about in the first place?

"Tweedledum," said Alice. "Tell me your views about the evolution of active systems like radio galaxies and quasars as the Universe grows older."

"It is quite remarkable," he said. "Active galaxies like quasars and radio galaxies were much more common in the past than they are now by a very large factor. When the Universe was about a quarter or a fifth its present age[11], these objects must have been about a thousand times more common than they are now[12]. We are living at a rather boring time in the Universe. Things must have been much more exciting when it was a bit younger, quasars going off all the time, radio jets zooming all over the Universe."

"But why did this happen?" asked Alice.

"To be perfectly honest, nobody really knows. I believe that it may be that galaxies as a whole only formed in the relatively recent past and you cannot have active galaxies until you have made the host galaxies first"[13].

"I disagree," said Tweedledee. "There is no reason why the galaxies should not have formed very early in the Universe. It may take time for big black holes to grow in the centers of active galaxies so that they can become quasars. Or maybe you don't see the quasars at great distances because they are obscured by dust."

The discussion was turning to one of the most interesting questions in cosmology, namely, how to find the most distant quasars in the Universe. The most distant quasars so far discovered emitted their light when the Universe was less than one-fifth its present age. It has, however, proved impossible so far to find any more distant quasars, despite a great deal of hard work by many astronomers with large telescopes. It might be that there were no quasars at earlier times.

Alice said, "Well, we know one thing for certain: The Space Telescope has maximum sensitivity for starlike objects and so it is the ideal instrument for detecting these very distant quasars. The only problem is how do you find the distant quasars among all the other objects which will appear in the ultra-deep images of the sky which will be taken with the Wide Field/Planetary Camera."

"Well," said Tweedledee, "the radio astronomers know exactly where to look because they have measured positions for their quasar candidates very

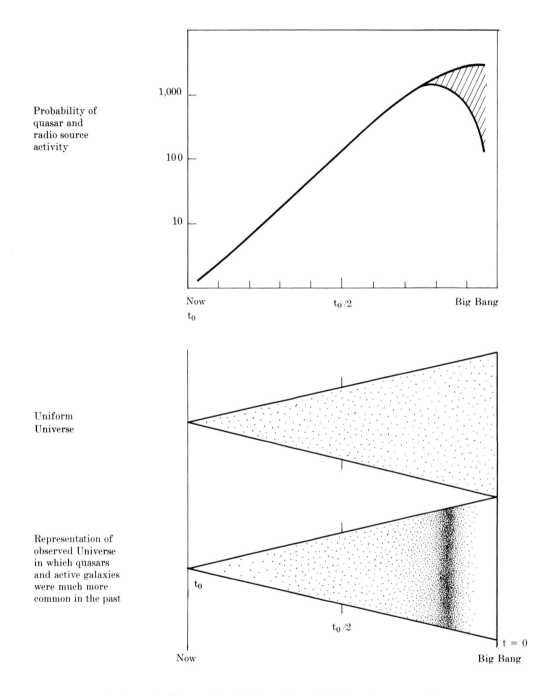

Probability of
quasar and
radio source
activity

Uniform
Universe

Representation of
observed Universe
in which quasars
and active galaxies
were much more
common in the past

A schematic diagram showing how the probability of quasar and
radio source activity has changed with cosmic time. There was
much more activity in the distant past. Representation of the
observed Universe in which quasars and active galaxies were
much more common in the past than they are at the present time.

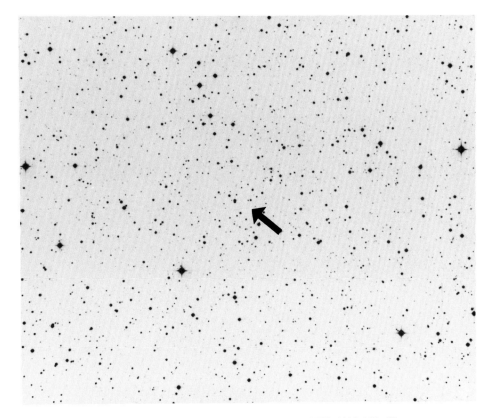

A photograph of the field of the quasar PKS 2000-330. The
arrow points to the quasar, which was discovered because it is a
strong source of radio waves. It emitted its light and radio waves
when the Universe was about one-fifth its present age.

accurately. In fact, if there are quasars which emitted their radiation when
the Universe was a twentieth its present age and if they are not obscured,
there is no reason why Space Telescope should not be able to observe them."

"These questions are related to the origin of galaxies, clusters, and other
large scale structures in the Universe[14], are they not?" asked Alice.

"Yes," said Tweedledum. "A late epoch of formation of galaxies is consistent with the observed large scale distribution of galaxies in the Universe.
According to the adiabatic model, big things like clusters and superclusters
of galaxies form first in the Universe and then galaxies form by fragmentation of these large-scale systems. This must happen late in the history of the
Universe. This theory also accounts naturally for the stringy structure which
we see in the distribution of galaxies and clusters. This is because the large
clouds collapse to form pancakes and so we expect to see flattened structures

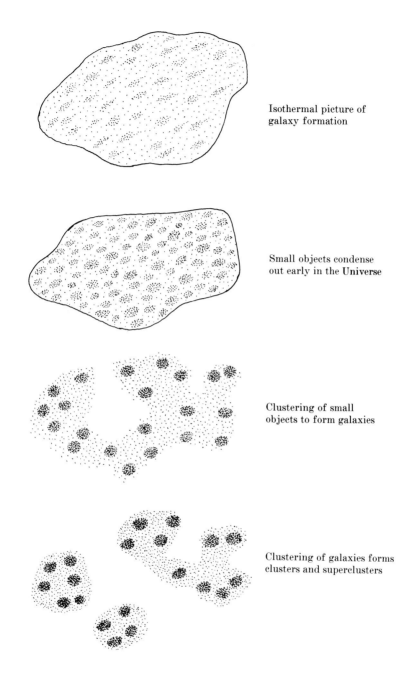

Isothermal picture of
galaxy formation

Small objects condense
out early in the Universe

Clustering of small
objects to form galaxies

Clustering of galaxies forms
clusters and superclusters

A schematic diagram illustrating the two main classes of theory
of the origin of structure in the Universe, the adiabatic and
isothermal theories of galaxy formation.

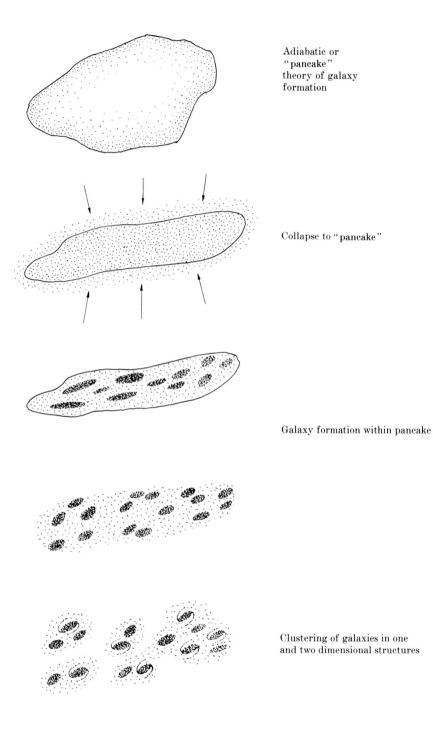

Adiabatic or
"pancake"
theory of galaxy
formation

Collapse to "pancake"

Galaxy formation within pancake

Clustering of galaxies in one
and two dimensional structures

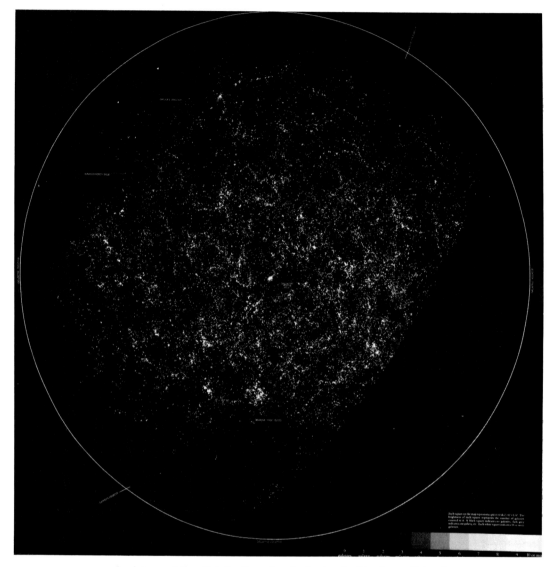

A picture of the distribution of galaxies in the Northern Galactic
Hemisphere. The center of the diagram corresponds to the view
looking vertically out of the plane of our Galaxy, and the outer
white circle corresponds to the view looking through the galactic
plane. The coordinates have been chosen so that equal surface
areas on the sky are properly represented. The lack of galaxies
towards the edge of the diagram is due to obscuring dust in the
plane of our Galaxy. The "bite" out of the diagram at the bottom
right occurs because this area of sky was not surveyed. The light
areas correspond to areas of high surface density of galaxies. The
bright "cluster" in the center of the diagram lies in the direction
of the Coma cluster.

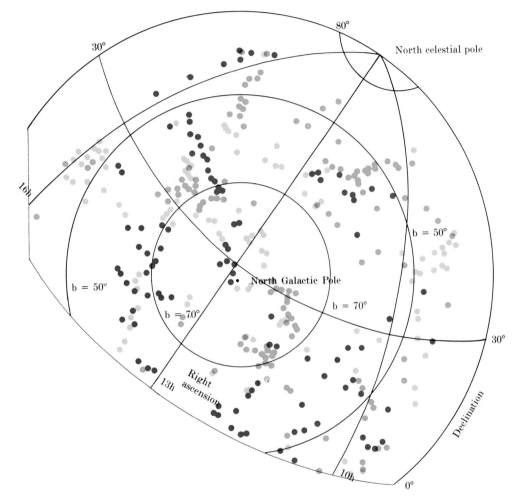

The distribution of bright galaxies in the direction of the North Galactic Pole. The blue dots correspond to galaxies with recession velocities in the range 3000 to 4000 km s^{-1}, the yellow dots to velocities in the range 4000 to 5000 km s^{-1}, and the red dots to velocities in the range 5000 to 6000 km s^{-1}.

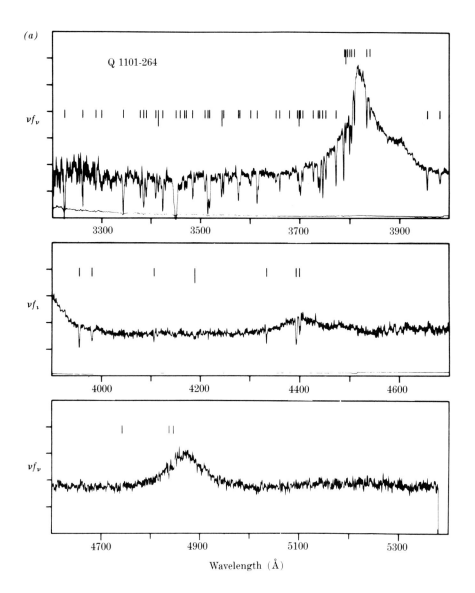

(a) The spectrum of the large redshift quasar Q 1101-264
showing the "forest" of absorption lines to the short wavelength
side of the redshifted Lyman-α line, which is observed at about
3800 Å. There are few absorption lines at wavelengths longer
than 3800 Å. (b) A schematic diagram illustrating the origin of
these absorption systems, which are caused by clouds of
intergalactic gas distributed along the line of sight to the Earth.
If observations are restricted to the visible waveband, only those
present in large redshift quasars can be observed. With the
Hubble Space Telescope, such features can be observed in low
redshift quasars in the ultraviolet waveband as well.

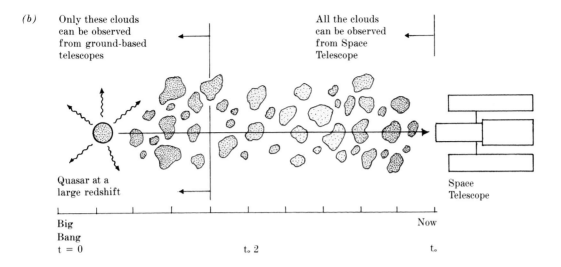

(b) Only these clouds can be observed from ground-based telescopes

All the clouds can be observed from Space Telescope

Quasar at a large redshift

Space Telescope

Big Bang
$t = 0$

$t_o 2$

Now

t_o

and large holes in the distribution of matter in the Universe, which is exactly what we see."

"Nonsense," said Tweedledee. "I can explain the origin of galaxies early in the Universe by the isothermal model, according to which the little objects form first and then cluster to form large objects like clusters and super-clusters. With this theory I can explain why even the most distant quasars seem to have the same abundances of the elements as nearby systems."

"It is very convenient to have discovered these very luminous objects, the quasars, at such great distances," said Alice. "We will be able to look for evidence of intergalactic gas from the absorption features which this gas would imprint upon the otherwise smooth spectrum of the quasar"[15].

"Quite right," said Tweedledum. "Many of these absorption features have already been seen. They are due to diffuse clouds of intergalactic matter containing the primordial abundance of hydrogen, helium, and the light elements. They are probably clouds which did not condense into galaxies and clusters when they first formed."

"Contrariwise," said Tweedledee. "We know that some of the absorption systems are not diffuse intergalactic clouds but are ejected from the quasars themselves. It is quite possible that a number of these features are simply clouds of gas ejected from quasars."

It was obvious that Tweedledum and Tweedledee were just itching to have another battle and Alice made one last attempt to prevent this happening. She couldn't help but notice, however, what a marvelous set of problems they were exposing, problems which could be uniquely addressed by observations with the Space Telescope.

"So far, you have only been talking about exotic systems at large distances. Wouldn't it be nicer to look at more normal things like ordinary galaxies at these large distances?" Alice asked.

"Yes, one can do that very well," said Tweedledum. "As you yourself pointed out in the last chapter, the typical deep image taken with the Wide Field/Planetary Camera will be an image of the Universe as it was when it was about half its present age."

"And what do you expect to see?" said Alice.

"Well, Professor Longair asked me to show you one of his diagrams just to demonstrate how well you can now do from the ground if one works very hard. This picture shows the Hubble diagram for radio galaxies observed by Simon Lilly and himself with the UK Infrared Telescope in Hawaii[16]. You can see that this diagram now stretches out to redshifts of 1.6, and the light from these systems is all ordinary starlight. You can see that all the galaxies at large redshifts are somewhat brighter than is expected for any of the standard world models. They attribute this to the evolution of the underly-

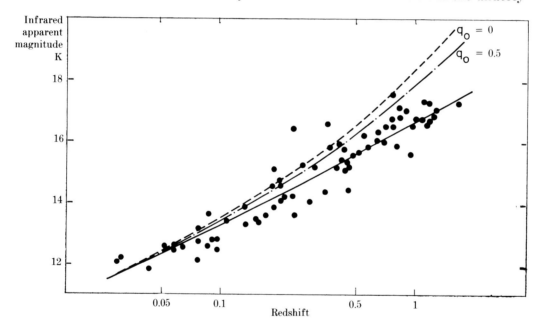

The redshift-magnitude relation observed in the infrared waveband for radio galaxies associated with powerful radio sources. The dashed lines indicate the expected relation if the luminosities of the radio galaxies have not changed with cosmic time and the deceleration parameter has values of 0 and 0.5. The solid line shows the expected relation taking into account the expected amount of evolution of the stars in these galaxies.

ing stellar population in these galaxies. They say that this is strong evidence that exactly the sort of evolution which we expect to see in the stellar population of galaxies is in fact occurring. The same sort of work can very easily be carried out on lots and lots of ordinary galaxies with the Space Telescope."

Alice was profoundly impressed, as she had to be since this was her author's work. She recalled him saying that the most significant thing about these observations was the fact that there exist and will exist facilities for studying the detailed astrophysics of the Universe at times much earlier than the present. This will change astrophysical cosmology from being merely a subject for theoretical speculation into a concrete subject constrained by real observation.

Alice looked back at her list of questions. "Well," she said. "We have only the origin of the Universe itself to deal with. Who's going to tackle that?"

"He can!" said Tweedledum.

"He can!" said Tweedledee.

"This is most unusual modesty for Wonderland people. Aren't you prepared to speculate just a little bit about the origin of the Universe?" asked Alice.

"We've got quite enough to argue about without thinking about that. You tell us your opinion," they replied.

Alice thought for a moment and said, "I like to believe that we probably understand the basic physics of the Universe from the time when it was about one one-hundred-thousandth of a second old right up to the present time. That is because we are using physics which has been tried and tested in the laboratory. However, earlier in the history of the Universe, it was so hot that we ran out of secure and tested physics. We have to be guided by the very best theory which the particle physicists have to offer and see how far we can get.

"We will have to wait and see how successful these theories are, but there are already some encouraging signs. We haven't really talked about the three basic questions of Big Bang cosmology [17] : (1) Why is the Universe made out of matter rather than an equal mix of matter and antimatter? (2) Where do the irregularities from which galaxies are formed come from? and (3) Why is the Universe so uniform on a very large scale?

"There are several aspects of modern theories of elementary particles which look promising as solutions to these problems. First of all, contrary to what one might expect, these theories predict that there is indeed a slight imbalance between matter and antimatter, just as we observe. Second, the theories may be able to account for the small irregularities in the early Universe. Third, because of the peculiar dynamics of the very early Uni-

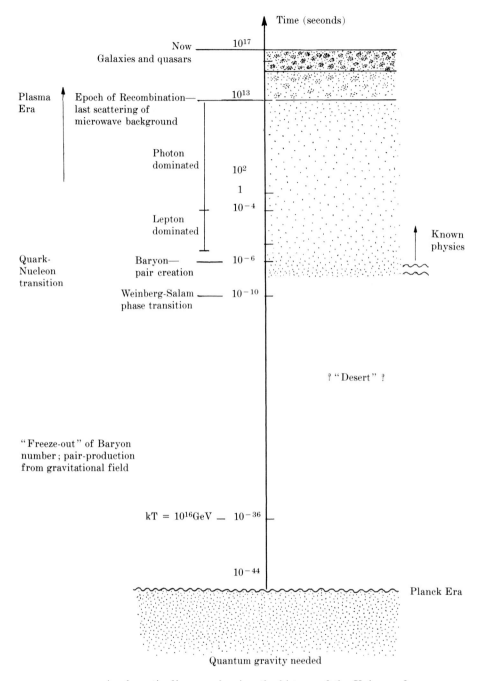

Time (seconds)

Now ——— 10^{17}
Galaxies and quasars

Plasma Era

Epoch of Recombination— 10^{13}
last scattering of
microwave background

Photon dominated 10^2
                           1
                           10^{-4}

Lepton dominated

Quark-Nucleon transition   Baryon— 10^{-6}
                                  pair creation

Weinberg-Salam ——— 10^{-10}
phase transition

Known physics

? "Desert" ?

"Freeze-out" of Baryon
number; pair-production
from gravitational field

$kT = 10^{16}\text{GeV}$ — 10^{-36}

10^{-44}

Planck Era

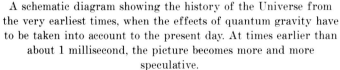

Quantum gravity needed

A schematic diagram showing the history of the Universe from
the very earliest times, when the effects of quantum gravity have
to be taken into account to the present day. At times earlier than
about 1 millisecond, the picture becomes more and more
speculative.

verse, we may be able to explain why the Universe looks the same in all directions.

"Ultimately, theorists will have to come to terms with the very earliest stages of the Hot Big Bang, when space-time itself ceases to have a definite meaning. We know that early enough in the Universe, when it was only about 10^{-43} seconds old, we have to consider the effects of quantum gravity and then we enter a Wonderland beyond anything we have encountered today. Space and time dissolve into a quantum foam and in the process, the point-like nature of the first instant of the Hot Big Bang itself may disappear.

"At least, that's what I think the theorists are saying," said Alice.

Tweedledum and Tweedledee looked at each other. "Now you tell me what Space Telescope will do to answer any of these questions!" said Tweedledum.

"And you tell me why I should take any of it seriously at all!" said Tweedledee.

Alice checked her list of questions again. "Well, there you are, Professor Longair," she said. "I hope that satisfies you."

"Clever chap, that Tweedledum. Must see if we can find a research fellowship for him"[18].

"Professor Longair, that's very unfair!" protested Alice. "You are just showing off your prejudices. Tweedledee's views are just as sound!"

Alice thought she now had quite enough material for her essays and was about to thank Tweedledum and Tweedledee for all their help when she saw that they were very angry with each other.

"You got it all wrong again, you twit!" said Tweedledum.

"You just don't understand the basic problems, you nincompoop!" said Tweedledee. "Let's have another battle."

This time Alice was not too worried because she remembered they had a battle every day and finished in time for tea. She helped them on with their armor and left them happily battling about the problems of cosmology.

Another Old Friend

Alice felt she now had plenty of material for her essays and yet she was still not happy about the very first one—"Why will the Space Telescope revolutionize mankind's understanding of the Universe?"

Just then she saw coming into view her best friend from Looking-glass Land, the White Knight[1]. He was riding his horse and falling off every 10 meters or so, just as before. She remembered that he was by far the kindest of all the creatures in Looking-glass Land and she was very fond of him.

"White Knight," she said. "I am so pleased to see you again. I have had very helpful discussions with all your friends here but I still have this one question to answer. Can you help me?"

"My goodness, Alice! How you have grown," said the White Knight. "Of course, let me try."

"I've got lots of material now but somehow I need to put the whole subject together. How would you do it?"

The White Knight looked much older than before, but he seemed much wiser too. "I don't know," he said. "I've been wrong far too often to try to give you any general conclusion at all. What I know is that things change and what seems to be the most important thing today will appear to be a minor issue to future generations of astronomers and cosmologists. How could anyone have guessed there would be strong radio jets coming out of radio galaxies? Who would have thought of trying to look for neutron stars at long radio wavelengths? Who would have guessed that a binary neutron star system would have provided the best test-bed for general relativity?"[2]

Alice waited while he remounted his horse.

"The only thing common to these great developments is the vision of dedicated scientists pushing back the frontiers of technology to tackle old problems in new ways. That is the fundamental significance of Space Telescope for mankind. For all of us, it is a monument to the living force within us which drives us on to probe and understand our Universe. It symbolizes

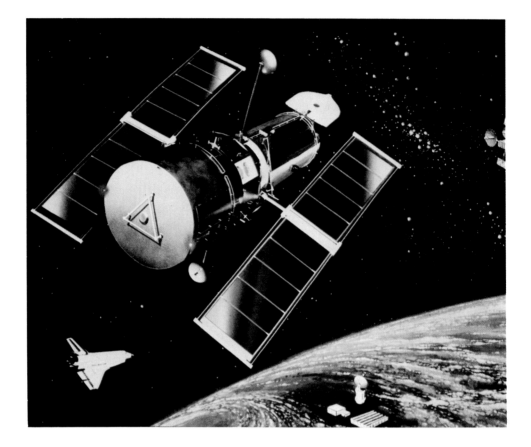

the act of faith by which all scientists pursue their research objectives. That faith is that by rational argument, by experiment, by observation and discovery, we can understand the physical nature of our Universe and that it is our duty as scientists to pursue these goals to their logical conclusions. The Space Telescope is perhaps the most significant symbol we have today of our society's belief in the importance of pure knowledge as part of our culture. We can only salute those who have made it possible and who have reinforced our belief in the pure cultural and intellectual significance of our scientific activity."

The White Knight rode off slowly. He fell off again twice before he disappeared from view. Alice shook herself and found she was still in the lecture room, but now it was empty and all the students had gone home. To her horror she realized that she had not been asleep at all.

The End

PART II *Annotations*

1. *An Old Friend*

1. Throughout the text, Alice's opinions and answers are those which might be expected of a very good student of astronomy and astrophysics in 1984. In my opinion, Alice would have obtained full marks for her performance in Wonderland and Looking-glass Land.

2. Humpty Dumpty first appears in *Through the Looking-glass* in chapter 6, "Humpty Dumpty."

3. The *Apollo 13* rescue and the in-orbit repair of the *Solar Maximum Mission* satellite are two remarkable feats of space engineering achieved by NASA, the United States National Aeronautics and Space Administration. In the case of *Apollo 13*, the third of the manned space flight missions to the Moon in April 1970, an oxygen tank exploded early on during the transit between Earth and the Moon, endangering the lives of the astronauts. By skillful use of resources and by using the full engineering capabilities of the spacecraft, the NASA engineers found a trajectory that enabled the crippled spacecraft to orbit the Moon and then come safely back to Earth.

   In the case of the *Solar Maximum Mission* (sometimes known as *Solar Max*), the satellite was launched in February 1980 but developed a serious malfunction after about nine and a half months in orbit. The satellite was captured by the Space Shuttle in April 1984, repaired by the astronauts, and sent back into orbit with a refurbished set of instruments. It has been working successfully ever since.

4. Riccardo Giacconi is Director of the Space Telescope Science Institute, which is located on the campus of the Johns Hopkins University, Baltimore. He is one of the pioneers of X-ray astronomy and was Project Scientist for the *Uhuru* and *Einstein* X-ray satellite observatories. Lyman Spitzer is Chairman of the Space Telescope Science Institute Council and the pioneer of optical and ultraviolet astronomy from space. Bob O'Dell was Project Scientist for the Space Telescope Project from the time of its approval in 1977 until 1983. Many distinguished scientists and advisory bodies have been associated with the project, thus ensuring that the very best science will be obtained with the Hubble Space Telescope. Advice has been obtained not only from optical and ultraviolet astronomers but also from radio, infrared, and X-ray astronomers and from theoretical astronomers. There has been input from the international community through the involvement in the project of the European Space Agency and through the participation of individual scientists.

5. The capital cost of the Hubble Space Telescope was about $1.25 billion at 1983/84 prices in June 1984. It is the most expensive purely astronomical satellite ever built.

2. *How to Build a Space Telescope*

1. It was most fortunate that Humpty Dumpty accompanied Alice to the building where all the animals were constructing the Space Telescope because it was he who had explained the meanings of the words of "Jabberwocky" to Alice (*Through the Looking-glass,* chap. 6). As he explained:

   A *borogove* is a "thin shabby-looking bird with its feathers sticking out all round—something like a live mop";

   A *mome rath* is "a sort of green pig" which has lost its way; and

   Toves "are something like badgers—they're something like lizards—and they're something like corkscrews . . . also they make their nests under sundials—also they live on cheese."

2. One aspect of the Hubble Space Telescope project which had to be mastered at the very outset was the use of acronyms and the jargon of space science, engineering, and management. In many cases, it turned out to be impossible to communicate without developing some expertise in the language. The examples quoted in the main text are mild, and my colleagues and I can vouch for whole lectures given entirely in acronyms and Space Telescope jargon. I am almost ashamed to admit that I developed very rapidly a certain fluency in this form of sub-English.

   Translations of the sentences Alice heard are as follows:

   "The astronomical instruments which will be mounted on the Space Telescope include, in addition to the Wide Field/Planetary Camera and the Fine Guidance Sensors, a Faint-Object Camera, a Faint-Object Spectrograph, a High-Resolution Spectrograph, and a High-Speed Photometer Polarimeter."

   "The signal from the Fine Guidance Sensors passes through the Scientific Instrument Command and Data-Handling System and the System Support Module, is transmitted via the Tracking and Data Relay Satellite System to the NASA communication ground station, and hence to the Payload Operations Control Center."

   "The computer programs for making a preliminary analysis of the scientific data obtained by the Space Telescope which can be used by General Observers are provided by the Instrument Definition Teams and by the Science Data Analysis System at the Space Telescope Science Institute."

   Although the distortion of the language is regrettable, one cannot deny that the economy of expression is impressive.

3. I fear I have added another "Jabberwocky" parody to the vast existing corpus. "Jabberwocky" is the most famous nonsense poem in the English language. In "Jabberwocky II," the animals explain the main features of the design of the Space Telescope. The following annotations explain some of the important scientific features of different aspects of the telescope and also the background to the items referred to in each verse.

4.   The Jabberwock itself is a fearsome monster with "jaws that bite" and "claws that catch!" As Humpty Dumpty explained, " 'Twas brillig" means "at four o'clock in the afternoon—the time when you begin *broiling* things for dinner."

It had long been the ambition of astronomers to launch a complete satellite observatory into space, and the concept of a space telescope dates back to the pioneering writings of Herman Oberth in the 1920s. In 1946, Professor Lyman Spitzer wrote a report entitled "Astronomical Advantages of an Extraterrestrial Observatory." A brief review of the history of the Space Telescope is given by Spitzer in his introductory remarks to the volume *Scientific Research with the Space Telescope* and also in the article "The Space Telescope Observatory," by John Bahcall and Bob O'Dell, in the same volume (eds. M. S. Longair and J. W. Warner [NASA CP-2111, 1979]).

There were many stages in the development of the Space Telescope before the final design configuration described here was reached. Concepts for a large optical telescope in space were started in the early 1960s, and in 1971 NASA began detailed studies of a 3-meter telescope known as the Large Space Telescope. Final approval for the Space Telescope project was given by the United States Congress in 1977 for the 2.4-meter telescope. Because of the decrease in diameter of the primary mirror, the word *large* was removed from its title. In 1983, the telescope was formally designated the Hubble Space Telescope, in honor of the great pioneer of observational extragalactic astronomy and cosmology, Edwin P. Hubble.

5.   The astronomers wanted as large a telescope as possible, but in the end the decision to launch and service the Space Telescope using the Space Shuttle determined the final configuration of the telescope. The complete satellite observatory had to fit into the cargo bay of the Shuttle, and although a 3-meter mirror could have been fitted in, the other equipment needed to convert the telescope into a complete satellite observatory was so massive that to put all the weight behind the primary mirror would have produced an unbalanced distribution of weight and too much power would have been needed to change the orientation of the telescope in space.

The solution was to decrease the size of the primary mirror so that the equipment sections could be located around the main mirror, thus producing a design which was much better balanced and for which far less power was required to maneuver the satellite in space (see also note [9], below).

6.   A Cassegrain telescope is a reflecting telescope in which the light is reflected from the large primary mirror onto a secondary mirror which then focuses the light through a hole in the primary mirror to a point behind that mirror. This is a great advantage for the Space Telescope because all the scientific instruments—cameras, spectrographs, and so on—can be conveniently located behind the primary mirror. The design of the telescope is such that these scientific instruments can be replaced in orbit.

A Ritchey-Chrétien design is a clever optical configuration for the surfaces of the primary and secondary mirrors in which the images produced by the telescope are in focus over as large an angle in the sky as possible. In fact, the field of view of the Space Telescope is quite small, the image remaining sharply in focus to about 10 arc

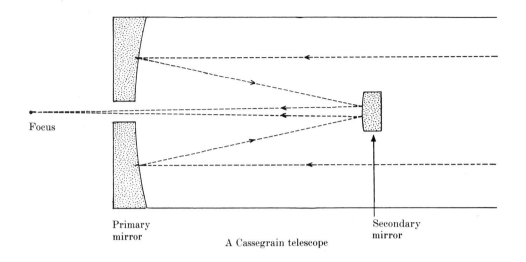

Focus

Primary mirror

Secondary mirror

A Cassegrain telescope

The layout of the mirrors and the optical path of a Cassegrain telescope. The light is focused behind the primary mirror.

minutes from the axis of the telescope; within that radius, however, the images will be as sharp as is theoretically possible for a mirror 2.4 meters in diameter for observations in the visible waveband.

Some of the specifications of the Hubble Space Telescope are given in Appendix 1.

7. One of the main reasons for sending a large telescope into space is that on the surface of the Earth we observe the Universe through the Earth's atmosphere. Even when it looks perfectly clear, we do not obtain the sharp images we expect to see according to the theory of optical telescopes because small fluctuations in the atmosphere blur the image we see to about 1 arc second in size. In fact, theoretically, the images we observe with a 5-meter telescope should be about 30 times smaller, or about 0.03 arc seconds in size. The theoretical limiting sharpness of the image of a telescope is called the diffraction limit. One of the major goals of the Space Telescope project has been to build a telescope for use in space which will produce this huge increase in sharpness over what we can obtain from the ground. According to the specifications of the Space Telescope, the images will be at least 10 times sharper than those obtained from the ground. In the ultraviolet waveband, where the wavelength of the radiation is shorter than in the optical waveband, the gain in sharpness will be even greater.

A second huge advantage of diffraction-limited optics in space is that, because much sharper images are obtained in space, the background light of the sky, which limits how faint one can see in deep exposures, is reduced by a large factor. Very much less unwanted background light falls on each picture element (or pixel) of the images taken with the Hubble Space Telescope (see p. 48).

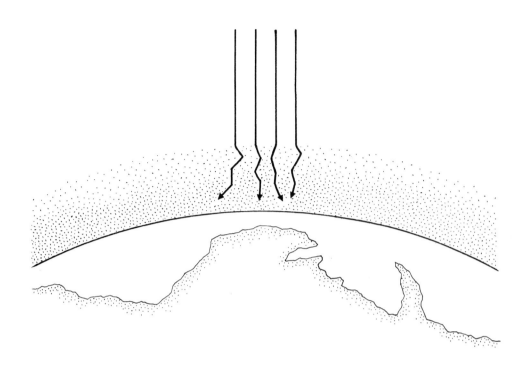

Illustration of the origin of the blurring of astronomical images
caused by irregularities in the Earth's atmosphere. This
phenomenon is known as "astronomical seeing."

8. The graphite-epoxy struts hold the primary and secondary mirrors apart. The great
    advantage of using graphite-epoxy materials is that they are very strong for their
    weight and they do not expand or contract with varying temperature, unlike
    materials such as steel or aluminum.

    As explained in note [5], the scientific instruments are located in the aft section of
    the telescope. These are huge instruments, each axial instrument being roughly the
    size and shape of a British telephone box (in fact, their dimensions are 0.9 × 0.9 ×
    2.2 meters^3). They are sometimes called coffins, but as such they would accommodate
    a large animal or person. There are six scientific instruments, five of which are
    provided by teams under the direction of a principal investigator; the sixth is the
    Fine Guidance System, which not only keeps the telescope pointing accurately at the
    correct point in the sky to a very high degree of precision but can also be used to
    measure very precisely the angular distances between stars and other pointlike
    objects. The five scientific instruments and their principal investigators are:

    The Wide Field/Planetary Camera—James A. Westphal
    The Faint-Object Camera—provided by the European Space Agency, project
        scientist—F. Duccio Macchetto

The Faint-Object Spectrograph—Richard J. Harms
The High-Resolution Spectrograph—John C. Brandt
The High-Speed Photometer/Polarimeter—Robert C. Bless

Some technical details of these instruments are given in Appendix 2.

9. As explained in note [5], a large part of the equipment was grouped around the main ring of the telescope to keep as much of the mass as close to the center of gravity as possible. This minimizes what is known as the moment of inertia of the satellite, which simply means its resistance to being maneuvered from one configuration to another.

10. Communication to Earth is provided by high-gain antennae on the satellite to the United States' *TDRSS* satellite, which has a high capacity for sending information to and from many satellites orbiting the Earth. *TDRSS* stands for Tracking and Data Relay Satellite System.

   The Space Telescope project was fully approved in 1977. The main contractors were the Perkin-Elmer Corporation, for the construction of the Optical Telescope Assembly (OTA), and the Lockheed Missile and Space Corporation, for the System Support Module (SSM). The SSM consists of all the structures and equipment which convert the telescope into a satellite. Besides the Faint-Object Camera, the European Space Agency is providing the solar array, which gives electrical power to the satellite.

11. Alice's summary of the key advantages of the Space Telescope are worthy of a little more explanation. Let us consider her points in order.

   a. The meaning of diffraction-limited optics has already been explained in note [7]. The primary mirror has been polished by the Perkin-Elmer Corporation to the highest accuracy that has ever been achieved in an astronomical mirror 2.4 meters in diameter. Thus, there is great confidence that the theoretical performance of a 2.4-meter mirror will be achieved in the optical waveband. The mirror seems so good that diffraction-limited performance is likely to be achieved at a wavelength three times shorter—that is, in the ultraviolet waveband, which is inaccessible from the surface of the Earth. This means that images even three times sharper than in the optical waveband are likely to be obtained. At even shorter wavelengths, it will be possible to obtain some information on an even finer scale by using special image-processing techniques.

   b. Besides the great advantage of very sharp images, placing the telescope in orbit will open up many wavebands which are inaccessible to observation from the surface of the Earth because these radiations are absorbed by the Earth's atmosphere (see p. 29). The telescope is designed so that it can focus all wavelengths of radiation longer than about 115 nanometers (1150 Angstrom). In practice, this means that the ultraviolet, optical, infrared, and submillimeter wavebands can be observed with the telescope. In fact, there are no instruments in the initial complement which can

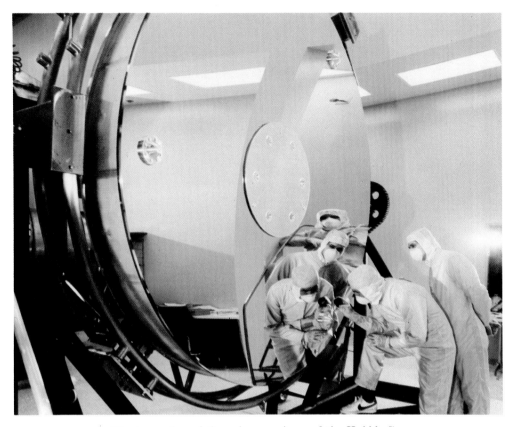

The inspection of the primary mirror of the Hubble Space
Telescope at the laboratories of the Perkin-Elmer Corporation.

detect radiation with wavelengths longer than 1 micron; in other words, the telescope will initially be an optical and ultraviolet telescope. However, there may well
be proposals to include infrared instruments on the telescope when the time comes to
replace some of the first-generation instruments.

c. The "weather" in space is not always perfect for astronomy for a number of
reasons. First of all, the telescope cannot be pointed anywhere on the sky. It is very
important at all stages of the mission to ensure that stray light does not enter the
sensitive instruments; if it does, they will fail to achieve their full sensitivity and
perhaps even be irreversibly damaged. This means that the telescope cannot point
too close to the Sun, Moon, or the bright Earth. Thus, only a limited range of targets
can be observed when the telescope is on the bright side of the Earth. Second, it is
important to maintain thermal balance throughout the telescope. Third, one must
maneuver so as to maintain the power supply to the telescope through the pho-
toelectric cells on the solar panels. Fourth, certain regions of the Earth's upper
atmosphere contain very high intensities of energetic particles, and these will cause
unacceptably high background radiation in the instruments. It is likely that some of

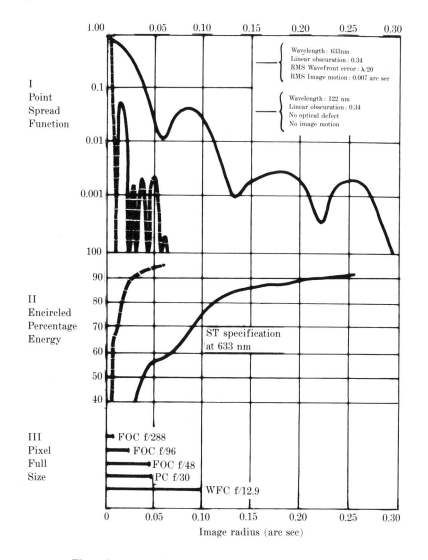

I
Point
Spread
Function

II
Encircled
Percentage
Energy

III
Pixel
Full
Size

Wavelength : 633nm
Linear obscuration : 0.34
RMS Wavefront error : λ/20
RMS Image motion : 0.007 arc sec

Wavelength : 122 nm
Linear obscuration : 0.34
No optical defect
No image motion

ST specification
at 633 nm

FOC f/288
FOC f/96
FOC f/48
PC f/30
WFC f/12.9

Image radius (arc sec)

The point spread function of a telescope shows how much of the
light of a point object like a star is contained within a given
angular radius of the star's position. At optical wavelengths, the
predicted performance of the Hubble Space Telescope will be
very close to the theoretical limiting angular resolving power of
a 2.4 m telescope. At ultraviolet wavelengths (about 200 nm), the
angular resolution is greater than at optical wavelengths by a
factor of about three. The top diagram shows the intensity
distribution of a point source; the center diagram shows the total
amount of energy received within a given angular radius of the
position of the star; the bottom diagram shows the pixel sizes of
the detectors in the cameras, showing that the cameras will be
able to take full advantage of the very high angular resolution of
the telescope.

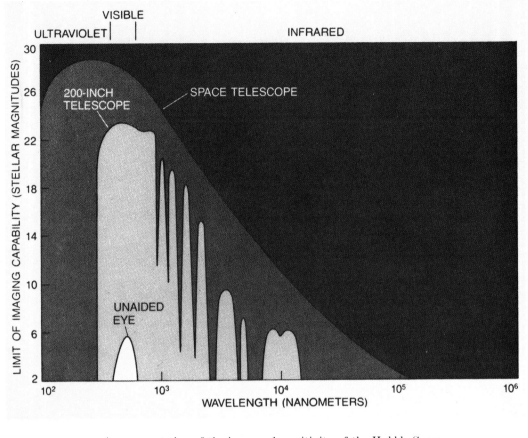

A representation of the improved sensitivity of the Hubble Space
Telescope for starlike objects compared with that of other
telescopes and satellites. The vertical axis is given in
astronomical magnitudes in which 2.5 units corresponds to a
factor of 10 in brightness.

the instruments will not be able to make observations when the telescope passes
through these regions. For all these reasons, the planning of observations with the
Space Telescope is a very complex matter. The constraints on stray light, for
example, are so severe that most observations will have to be taken when the satellite
is on the dark side of the Earth. In terms of operational efficiency, the telescope will
probably observe astronomical objects for about 35 percent of the total time. During
this period, observing conditions should be perfect.

d. and *e.* The Hubble Space Telescope will carry a full complement of instruments
as listed in note [8] and Appendix 2. These are still among the very best instruments
that could be built, although it must be remembered that their designs were frozen in
1977.

A model of the Hubble Space Telescope being serviced by the
Space Shuttle.

f. One of the specifications of the project is that the telescope will have a 15-year lifetime in space. In the original concept of service and refurbishment by the Space Shuttle, it was envisaged that the telescope would be visited once every two and a half years by the Shuttle and brought back to Earth every five years for a major refurbishment. Thus, to achieve a 15-year lifetime in space, there would have to be an initial launch and two subsequent flights of the telescope. The status of these proposals is currently under review. Most scientists are of the view that it will be the most productive scientifically to keep the telescope in orbit as long as possible and to make changes to the instruments in orbit. Decisions on these policies have yet to be made.

12. The initial launch date was to have been October 1983, but various technical and funding problems delayed the launch. As a result of the *Challenger* disaster, in January 1986, the Shuttle program was suspended. The expected launch date of the Hubble Space Telescope is now 11 December 1989.

3. *The Planetary Caucus-race*

1. The Space Telescope Science Institute is responsible for organizing the way in which astronomers make observations with the Hubble Space Telescope and then convert the new data to a form in which the astronomers can undertake scientific analyses of their observations. The Institute is located on the campus of the Johns Hopkins University in Baltimore, Maryland. This location was selected by competition. AURA, the Association of Universities for Research in Astronomy, had proposed that Johns Hopkins act as host institution. The prime function of the Institute is to provide services to the whole astronomical community in relation to all aspects of the interaction of astronomers with the Space Telescope project. To carry out this task effectively, there is a significant research effort by the astronomers at the Institute amounting to about 10 to 15 percent of the total effort. This is essential so that members of the Institute can respond effectively and as scientific equals to the many demands which will be placed upon the Institute by the scientific community. The Institute employs a total staff of about two hundred fifty persons.

2. Alice meets the Red Queen first in chapter 1 of *Through the Looking-glass*. The Red Queen is a rather brusque character and is always dashing everywhere just like the queen in a chess game.

3. The American Astronomical Society is the most important organization for professional astronomers in the United States. Regular meetings take place throughout the country, and these are very important for the dissemination of ideas and new results. The Baltimore meeting at which the lecture "Alice and the Space Telescope" was presented was attended by more than a thousand astronomers, and the scientific papers were delivered in seven parallel sessions.

4. The Caucus-race and the birds and animals who appear in this chapter meet Alice in chapter 3 of *Alice's Adventures in Wonderland*, "A Caucus-race and a Long Tale."

5. Although the study of the Solar System and the planets is plainly an astronomical subject, it differs from extra–Solar System astronomy in that, because of the development of space vehicles, it is now possible to visit the planets or at least to fly very close to them. This has changed the types of question which can be asked. The planets are now the subject of intense study by meteorologists, plasma physicists, and geologists as well as astronomers. Thus the traditional astronomy community has been expanded to include geologists, geophysicists, meteorologists, environmental scientists, and so on. Space probes can investigate in detail the environments of the planets and the material in the space between the planets, the interplanetary

medium, and the Solar Wind. In these environments, plasma physicists can study the behavior of very low density plasmas by directly measuring particle concentrations, the velocities of the particles, and the strength of the interplanetary magnetic field.

These developments have brought different types of scientist into the study of the Solar System, and many new and important scientific disciplines have developed, disciplines in which the scientists pursue quite different interests from the traditional interests of astronomers.

6. For details of the Wide Field/Planetary Camera, see Appendix 2.

7. *Voyager 1* and *Voyager 2* have been among the most spectacular and successful space missions dedicated to the study of the planets. The two space vehicles, launched at an interval of four months, set out towards Jupiter and the outer planets. *Voyager 1* reached its point of closest approach to Jupiter on 5 March 1979 and *Voyager 2* on 9 July 1979. *Voyager 1* was placed in a trajectory that passed close to Uranus and then will carry it out of the Solar System. *Voyager 2* also flew on to Saturn and was then placed in a trajectory which will pass close to Uranus on 24 January 1986. Then, if the spacecraft is still functioning, it will pass by Neptune in late August 1989.

The image shown on page 16 shows Jupiter as it was observed at a distance of 2.5 million km from the planet by *Voyager*. The wealth of detail which can be observed in this picture is typical of the quality of the images which will be taken regularly by the Hubble Space Telescope.

8. The Galileo project is a NASA program to send a large space probe into orbit around Jupiter. The *Galileo* space vehicle is scheduled to orbit Jupiter for at least 20 months in a trajectory that will bring it close to all of Jupiter's large moons, which are known as the Galilean satellites. They are named after Galileo Galilei, who first observed them through one of his early astronomical telescopes. They are, in increasing order of distance from Jupiter, Io, Europa, Ganymede, and Callisto.

There will be many experiments on board the *Galileo* spacecraft, including one in which a spacecraft will be dropped into Jupiter's atmosphere. The cameras on *Galileo* will be so close to the surface of the planet that they will take extremely high-definition pictures of the surfaces and atmospheres of the planet and its satellites. However, these sharp images will only be obtained over a very small area of the planet's surface. Simultaneous observations with the Hubble Space Telescope will make it possible to relate the small-scale phenomena observed by *Galileo* to the large-scale properties of Jupiter's atmosphere as observed by the Space Telescope.

9. Io, the innermost of the Galilean satellites, is one of the most exotic objects in the Solar System. It is red-orange in color because its surface is largely covered with sulphur. The sulphur is probably ejected in the spectacular volcanic events which were observed on Io's surface during the *Voyager* fly-bys. There was continuous volcanic activity on Io during the *Voyager* encounter (see p. 17). No impact craters are visible on the surface of Io, because it is constantly being covered by debris from the volcanoes. This violent volcanic activity results from the continuous input of

Io

Europa

Ganymede

Callisto

A collage showing the very different surface properties of the
four large Galilean satellites of Jupiter. In order of increasing
distance from the planet, they are Io, Europa, Ganymede, and
Callisto.

energy into the interior of the satellite caused by tidal torques exerted by Jupiter
upon Io. The friction generated in the interior of the satellite by these tides has been
shown to be sufficient to melt a large fraction of the interior of Io. The volcanoes
therefore represent points where this fluid interior bursts through the rather thin
crust of the satellite.

In addition to these remarkable properties, Io is embedded in the environment of
Jupiter, which possesses a strong magnetic field. As Io passes through the magnetic
field, it causes electric and magnetic disturbances which result in the emission of low
frequency radio waves from Jupiter's environment. Io acts like an electrical conduc-
tor and thus is similar to an electric dynamo in Jupiter's magnetic field.

An image of the outermost rings of Saturn showing the twisting
of the outer ring. This is likely to show significant changes
during the lifetime of the Hubble Space Telescope.

10. Saturn has already been studied in detail by the *Voyager* spacecraft, but it took
images of the planet only over a short time interval. With the Hubble Space
Telescope, the long-term behavior of the atmosphere all over the planet can be
studied. Uranus and Neptune have very small angular sizes, the angles they subtend
on the sky being typically only 3.5 and 2.0 arc seconds, respectively. It is therefore
impossible to obtain any detailed information about the weather patterns on their
surfaces from observations from the ground. The large improvement in angular
resolution which will be possible with the Space Telescope will enable these studies to
be undertaken.

11. One of the most remarkable discoveries of the Voyager program was that the famous
rings of Saturn were each resolved into many finer rings (see p. 18). This seems
remarkable at first sight because we might expect the random motions introduced by
the gravitational influence of nearby bodies to distribute the matter of the rings

rather smoothly around the planet. In fact, the rings exist because of gravitational resonance effects associated with the presence of small satellites which lie within the rings of Saturn.

Some of the rings seem to be distorted, but these distortions cannot last very long dynamically. They may change over the time scales which will be accessible from the Space Telescope. This would help us understand the dynamical processes which allow the beautiful ring structures to persist for so long.

12. Pluto, the outermost planet of the Solar System, is also the smallest of the planets, and no images of its surface have been obtained from ground-based observations. The technique of speckle interferometry has been used to estimate its angular diameter, which is about 0.1 arc seconds (about 3,000 km). Thus, even with the Space Telescope, it will be difficult to resolve details of Pluto's surface, but it will be possible to measure its shape and also to distinguish its small satellite, Charon, as a separate body.

13. The surface and atmosphere of Mars were studied in great detail by the two *Viking* spacecraft which reached Mars in 1976. Each released a vehicle which landed on the surface, took photographs of its surroundings, and analyzed soil samples. Part of each spacecraft remained in orbit and took very high resolution pictures of the surface of the planet.

One of the *Viking Lander* images of the surface of Mars taken in the vicinity of the space vehicle's landing. The red color of the surface is due to the dust which covers much of the surface of the planet.

These photographs have produced remarkable evidence for dust storms, which must have been a major influence in determining the surface properties of Mars. The studies of these violent Martian storms may help us to understand what could happen to our own planet if a large amount of dust were liberated into our atmosphere. Such sources of dust on Earth include large volcanic eruptions or the impact of large meteorites on the Earth's surface. The aftermath of a major nuclear explosion would probably produce the same types of effect.

14. The Earth's atmosphere is a very complex system and the processes which maintain the rather delicate balance which enables it to exist are not yet fully understood. It is plainly of central importance to the continued existence of life on Earth that we do not upset this delicate balance by ejecting into the atmosphere chemical compounds which would change dramatically its chemical composition.

15. According to a wide range of theories, the small bodies in the Solar System, such as asteroids and meteorites, may well be the oldest "fossils" we can study to determine the original materials out of which the planets and the Sun itself were first formed about 4.6 billion years ago. The surfaces of the planets have changed greatly from their original state, whereas the small bodies have probably remained much as they were when the Solar System first came into being. Studies of these objects thus may provide us with clues to the sequence of events which took place when the Sun, the planets, and all the other objects of the Solar System were condensed out of the primitive solar nebula.

16. The comets are certainly primitive bodies in some sense, although their origin is far from clear. An interesting general point is that they may bring with them from interstellar space material whose chemical composition and properties are the same as those from which the Sun and the planets were formed. An example of special importance is the abundance of deuterium in comets. Since deuterium is a very fragile nucleus, it is readily destroyed inside stars; hence, it is a tracer of the nucleosynthetic history of the material of which the comet is composed. Observations of deuterium in comets will be one of the key projects to be undertaken by the High Resolution Spectrograph on the Hubble Space Telescope.

17. "The Mouse's Tale" is one of the most famous of the poems in *Alice's Adventures in Wonderland* (chap. 3). In the present version of the poem, the Mouse is lamenting the fact that the Space Telescope will not be able to observe Halley's Comet until well after the time when the comet passes closest to the Earth. This was forced upon the Space Telescope project by certain unavoidable delays in the program. Halley's Comet is of special importance because it exhibits the full range of phenomena observed in comets.

18. The animals list a few of the arguments sometimes put forward to indicate the importance of particular aspects of astronomical and Solar System research because

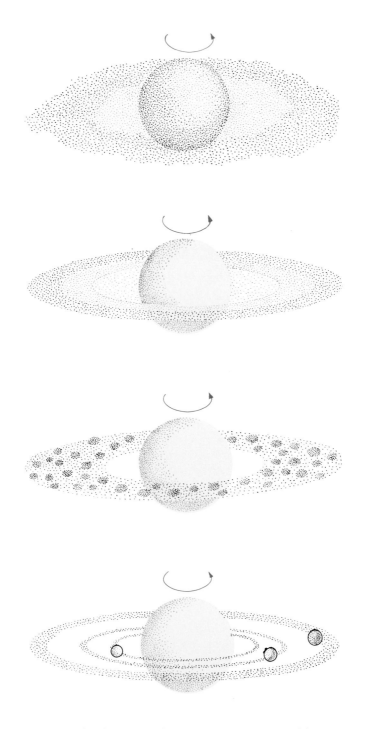

A schematic picture showing a sequence of events which may
have led to the formation of the Solar System out of the
primaeval solar nebula.

of their potential practical value to mankind. The case of the Martian dust storms and the understanding of the aftereffects of large releases of dust into the atmosphere is an excellent example of this type of argument. The case for the study of asteroids and the possibility that one of them might crash into the Earth is a serious one because such an event could have catastrophic consequences for life on Earth. One of the more plausible pictures for the extinction of the dinosaurs is that the event is associated with the cooling of the Earth's climate which in turn was associated with the release of a large amount of dust into the atmosphere following the collision of an asteroid with the Earth. If we were able to predict such an event, we might be able to take some action to avoid its worst effects and possibly even to avert it. The idea that black holes might one day provide us with an energy source to solve the energy crisis is a desperate attempt to make the study of these objects relevant to everyday life. No one takes this idea seriously at the present time.

4. *"Twinkle, Twinkle, Little Star"*

1. STARS AND STELLAR EVOLUTION

A little of the background to our present understanding of stars and stellar evolution may help explain some of Alice's questions and concerns.

It is convenient to consider the life of a star in three parts—its birth, life, and death. We will describe the birth of stars in more detail in the next chapter; here we will only treat the main phase of the life of a star and the ways in which stars ultimately die.

Astronomers can measure directly the surface properties of stars, and two of the most important of these are the total amount of light emitted by a star (its luminosity) and the temperature of its surface. What makes the study of the structure and evolution of stars one of the most exact of the astrophysical sciences is the fact that not all possible combinations of temperature and luminosity are found among the stars we see. If we plot quantities which are more or less equivalent to luminosity and temperature against one another, it is found that the stars occupy quite specific regions of this luminosity-temperature diagram. This plot is known as a Hertzsprung-Russell (or H-R) diagram or as a color-magnitude diagram. The shaded regions of the diagram show where virtually all the stars we know of lie. Most stars lie along the band known as the *main sequence*, which runs from top left to bottom right of the diagram. What distinguishes stars along this sequence is their mass. The most massive stars are at the top left of the main sequence, and stars of the lowest at the bottom right. Our Sun lies near the middle of the sequence and is a very ordinary star.

Extending from about the location of the Sun on the luminosity-temperature diagram up toward the top right is what is known as the *giant branch*. The stars in this region are large and cool. Extending across the diagram from the giant branch are stars of the *horizontal branch*. There are in addition stars which lie significantly to the left and below the main sequence; these are very faint, compact stars known as *white dwarfs*.

One of the main goals of the theory of stellar structure and evolution is to understand why stars appear only in certain regions of this luminosity-temperature diagram and how they evolve from one part to another.

We consider first of all the main sequence. The source of energy is nuclear reactions occurring in their centers. The most common element in the Universe is hydrogen, which is the lightest of all the chemical elements; the next heavier stable element is helium, which is also the next most abundant, having an abundance of about 25 percent by mass. All the heavier elements, including carbon, nitrogen, oxygen, and iron, amount to only about 1 to 2 percent by mass of all the elements. The centers of stars are so hot that hydrogen can be fused into helium, thereby liberating

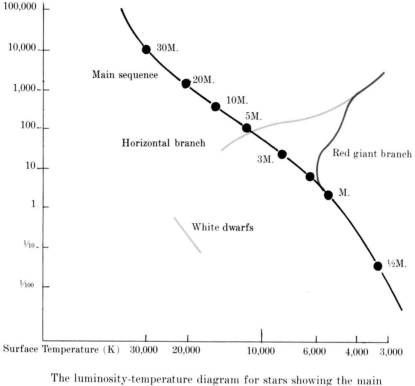

The luminosity-temperature diagram for stars showing the main
regions in which they are found. The names of the different
branches are shown. The luminosities of the stars are given in
units of the luminosity of the Sun $L_\odot$. The masses of stars along
the main sequence are shown in terms of the mass of the Sun
$M_\odot$.

the energy which is the origin of the light in all stars on the main sequence.
Theoretical models of stars using the best nuclear physics available from laboratory
measurements can give a convincing quantitative explanation of the observed properties of main-sequence stars. For most of the star's lifetime, it remains at one point
on the main sequence. A star like the Sun remains on the main sequence for about 10
billion years.

When a star has converted about 12 percent of its hydrogen into helium in its core,
it becomes unstable. The core contracts and the envelope of the star expands forming
a giant star. During this process, the position of the star on the H-R diagram moves to
the right and then up the giant branch. The instability involves a number of changes
in its internal structure. As the star collapses to higher densities and temperatures,
different nuclear reactions take place which fuel the star during these phases. All
stages after the star moves off the main sequence take place very much more rapidly
than its evolution on the main sequence. The star undergoes other instabilities
during which it sheds mass from its outer layers. When this occurs, the star moves
across to the horizontal branch and then evolves back towards the giant branch.

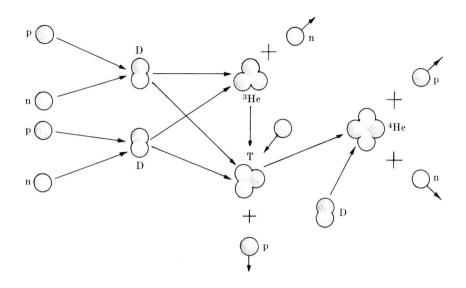

The synthesis of helium nuclei from protons and neutrons. The
reaction chains show the most common routes for the synthesis of
helium in the early stages of the Hot Big Bang.

During this evolution on the giant and horizontal branches, the helium in the star's core is burned into heavier elements, such as carbon, nitrogen, and oxygen. This is the origin of many of the chemical elements. After one or more excursions of this type, the star eventually ends up at the top right of the giant branch, an area occupied by long-period variable and unstable stars. When the star becomes unstable, the outer layers are blown off, producing the characteristic "planetary nebula" phase of the star's evolution, while the core of the star collapses to a white dwarf star and its position on the H-R diagram ends up to the lower left of the main sequence. This evolution is shown schematically on page 117.

This is the sequence of events followed by stars with masses about that of the Sun. More massive stars evolve much more rapidly, and it is likely that stars which are more than about four times the mass of the Sun do not end up forming white dwarfs but end their stellar lives as supernovae—exploding stars. What stellar remnant is left after these events is not certain, but we do know rather precisely what sorts of objects can be produced at the end of a star's lifetime.

There are three types of "dead star." In all three cases nuclear reactions no longer generate energy in the center of the star. One form of dead star is a white dwarf in which the star is held up by the pressure of very high density electron gas, known technically as electron degeneracy pressure. Typically, one cubic centimeter of white dwarf matter weighs about 1000 kilograms (or roughly one ton). The mass of a white dwarf is similar to that of the Sun, or less.

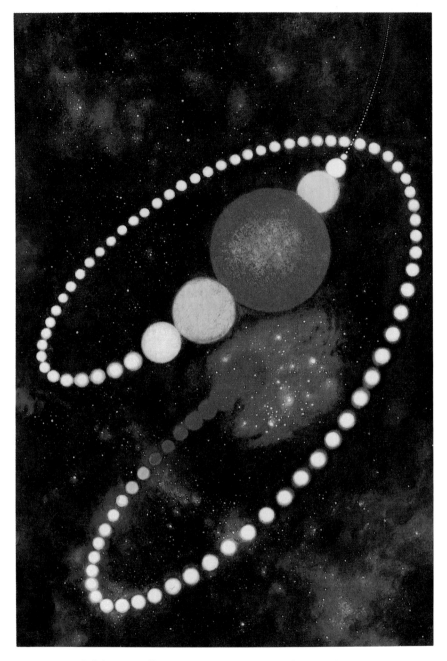

A "time-lapse" picture of the evolution of a star, from its
formation in a giant molecular cloud, through the main part of
its lifetime as a main sequence star, and ending in its expansion
to a red giant and subsequent existence as some form of dead
star. The interval between images, or "frames," of the star is
roughly one million years.

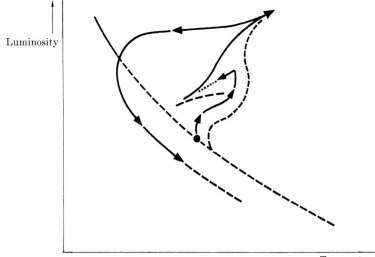

A schematic diagram showing the evolution of a star of
approximately the same or twice the mass of the Sun, from the
main sequence, up the giant branch, onto the horizontal branch,
to the peak of the giant branch, and finally through the
planetary nebula phase to a white dwarf.

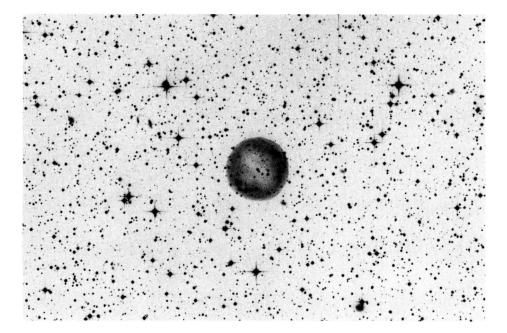

The planetary nebula 286 + 11.1. The slowing, expanding shell
of hot, glowing gas is excited by the intense, ultraviolet light of
the small, very hot star in the center.

(a)

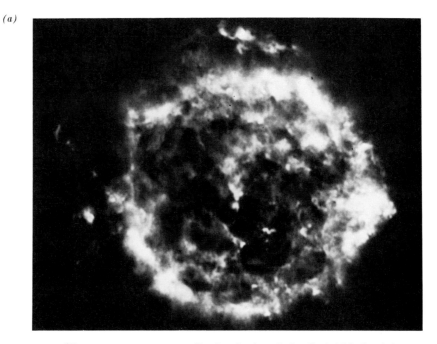

The supernova remnants Cassiopeia A and the Crab Nebula. *(a)*
The image of Cassiopeia A is a "radio photograph" made with
the Cambridge 5-km telescope. The shell-like structure is the
result of the explosion of the progenitor star, which must have
taken place about 250 years ago. *(b)* The Crab Nebula has a
different appearance. It exploded in 1054 and was recorded by
Chinese astronomers. Its appearance is due to a young pulsar at
the center of the nebula which acts as a powerful, continuous
source of high-energy particles and magnetic fields.

(b)

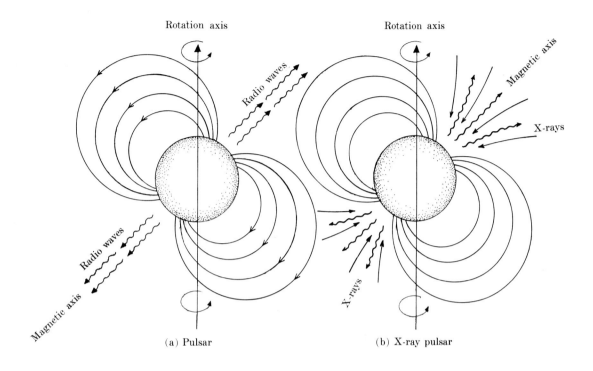

Models illustrating how magnetized rotating neutron stars can be sources of pulsed radio and X-ray emission. In both cases the magnetic axis of the neutron star is not aligned with the rotation axis. Beamed emission from the magnetic poles results in the observation of pulsed emission by a distant observer. In the case of the radio pulsar *(a)*, the radio emission is probably due to streams of particles ejected from the magnetic poles streaming out along the magnetic field lines. In the case of the X-ray pulsar, *(b)*, the X-ray emission is due to infalling gas being funneled down the magnetic poles and heated up in the accretion process.

A second possible end point is a *neutron star*. In this case the star is held up by the pressure of very high density neutrons and protons, or neutron degeneracy pressure. These stars have masses roughly the same as that of the Sun and are very compact indeed, having radii of about 10 kilometers. One cubic centimeter of neutron star material typically weighs about 1000 billion kilograms (or roughly a billion tons). Neutron stars have been found in radio pulsars and binary X-ray sources. Radio pulsars, which are rotating magnetized neutron stars, are observed to emit very intense pulses of radio emission once per rotation period, which is about one second. In binary X-ray sources the X-rays are produced by matter from the primary star falling onto the very compact neutron star.

The third possibility is that the star collapses to a *black hole* (see Chap. 6). If the collapsing star has mass greater than about two and a half times the mass of the Sun, a black hole is expected to form unless the collapsing star can find some way of losing mass and a stable neutron star can form.

Alice's list of questions can now be put in context.

a. Why are there too few solar neutrinos?

Alice explains this problem later in the main text. The basic point is that when we make observations of stars we obtain direct information about their surface properties only; we have to infer their internal structures from theoretical studies which turn out to be remarkably successful in explaining the observed properties of main sequence stars and their masses. It would be very nice, however, to have some direct means of testing the theory of the internal structure of the stars. This can be done by searching for the neutrinos which must be liberated in the nuclear reactions in the cores of stars. The neutrinos have virtually no interaction with matter, which means

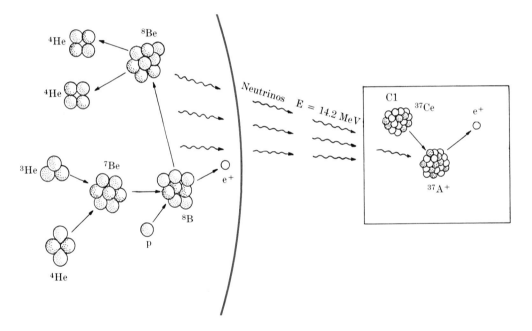

Illustration of the nuclear processes by which neutrinos have been detected from the Sun. The neutrinos which have been detected are produced in the rather rare reaction network shown in the diagram, the process which produces the neutrinos being the beta decay of boron-8 nuclei. These are detected on Earth by the nuclear transmutation they induce in chlorine-37 nuclei, converting them into argon-37. The neutrino flux from the Sun is found by measuring the amount of argon produced in a vast tank of perchlorothylene buried deep in the Homestake gold mine.

that they can escape from the very center of the Sun, but it also means that they are extremely difficult to detect on Earth. Although only a few are detected each year, solar neutrinos have been detected, but at a somewhat lower arrival rate than is expected theoretically. It is not clear at this stage just how severe this problem is for stellar astrophysics and for nuclear physics.

b. How do single and double stars and clusters of stars evolve?

We have outlined above how we expect single stars to evolve. Although evolution on the main sequence is quite well understood, the subsequent stages require much improved data to determine the stars' evolution more precisely. Although the discussion above concentrated upon single stars, we know that a large fraction of all stars are in fact double. This modifies a number of aspects of the evolution of stars, particularly the more massive ones. Many exciting new possibilities come about. Most of these phenomena are associated with the transfer of mass between one star and the other. This is the process which gives rise to the phenomena of novae and is also the power source for the high energy phenomena seen in binary X-ray sources.

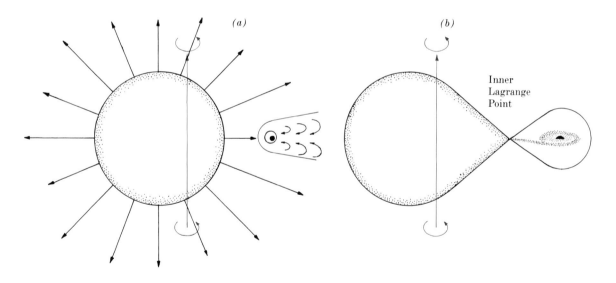

Diagrams illustrating two possible ways in which accretion of matter can take place by mass transfer in binary stellar systems. In case (a), the compact secondary star lies within a strong stellar wind and accretion takes place within a shocked cavity. In case (b), the surface of the primary star is distorted by the strong gravitational field of the secondary, and matter is drawn off the primary through the "Lagrangian point" to the secondary star. Accretion of matter onto the compact secondary star is expected in both cases, and this can generate intense X-ray emission, as shown on page 119 or through the formation of an accretion disk (see Chap. A7, note [4]).

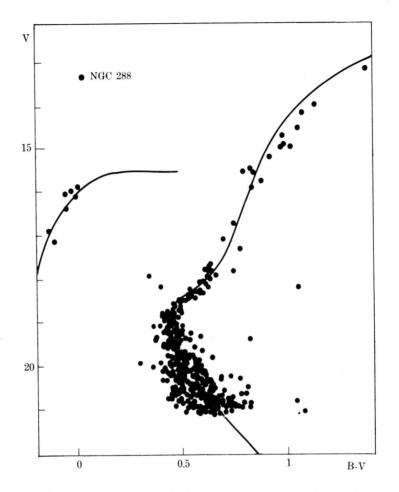

The observed color-magnitude diagram for the globular cluster
NGC 288. It is likely that a considerable part of the scatter about
the mean line is due to observational uncertainties.

Star clusters provide some of the best tests of the theory of stellar evolution.
Because all the stars can safely be assumed to have the same age, one can obtain a
"snapshot" of the state of evolution of stars of different masses in the cluster. The
luminosity-temperature diagrams for clusters provide some of the best information
about stellar evolution. In addition, star clusters, in particular the globular clusters,
provide an excellent testing ground for the theory of the dynamical evolution of
clusters of stars, that is, how the distribution of stars in the cluster evolves with time.
It is expected that, as a cluster evolves, some stars will acquire high velocities
through gravitational interactions, and be ejected from the cluster. As a result, the
cluster as a whole will contract. Studies of the detailed structure of clusters will
enable the theory to be tested, and insight will be gained into the ultimate fate of
clusters of stars.

c. How do stars lose mass?

We have indicated the various stages at which stars are expected to lose mass. Mass loss can be measured very effectively by observations in the ultraviolet region of the spectrum, as has been amply demonstrated by the *International Ultraviolet Explorer*. The profiles of emission lines from the material flowing out from stars have a characteristic shape, and these have been observed in many classes of star in the ultraviolet waveband (see p. 124). The processes of mass loss are very important because this is the means by which the elements synthesized in stars can be circulated to the general interstellar medium, out of which the next generations of stars will be formed. The mass loss may be quiescent, as in the case of our own Sun; or it may be more violent, when shells of material are expelled from the surface; or it may be extremely violent, when the whole star explodes as in a supernova explosion. All these processes contribute to the recirculation of the elements through the interstellar gas (see Chap. 5).

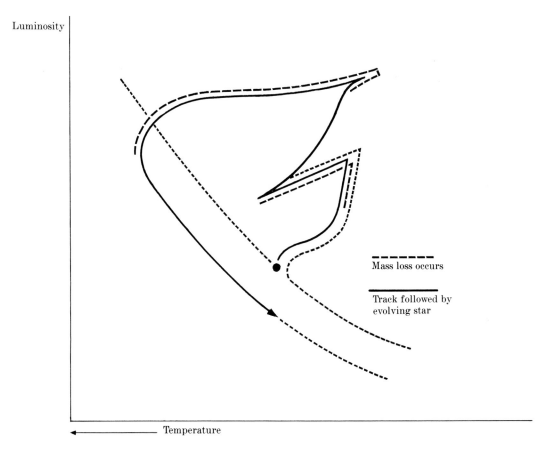

A schematic diagram illustrating the various stages in the evolution of a star during which it is likely to lose mass.

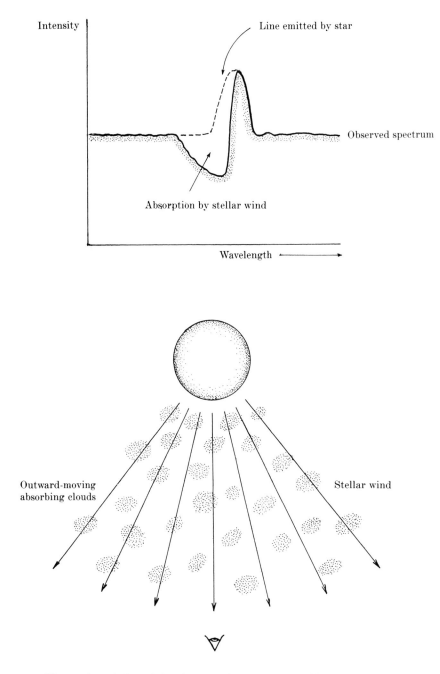

Illustration of the origin of the "P-Cygni" type profile of lines in
a stellar spectrum. Ions and atoms in the outflowing wind absorb
the line-radiation emitted close to the star and cause the
characteristic shape of the line profile. Profiles such as these are
observed in many types of star and are evidence of mass loss by
stellar winds.

d. How do stars die, and which become white dwarfs, neutron stars, and black holes?

Astrophysicists are very confident that these are the only three possible end points for stars : white dwarfs, neutron stars, or black holes. Many white dwarfs are known—they are hot, faint stars. Many neutron stars are now known from pulsar surveys and from the study of X-ray binaries. There are no absolutely certain detections of black holes, but there are several highly suggestive cases of dark companions in X-ray binaries and compact massive objects in active galactic nuclei. Most astrophysicists are confident about the existence of black holes, and these objects are now very much part of the everyday furniture of theoretical astronomy.

The big astrophysical problem is that it is not at all clear which types of star evolve into these different types of object. It is almost certainly the most massive stars which evolve into black holes because if the collapsing star has mass less than about twice the mass of the Sun, it can exist in a stable state as a neutron star or a white dwarf ; collapse to a black hole would not be mandatory. However, if the mass of the remnant is greater than this, collapse to a black hole is inevitable. What is very uncertain is how effective the processes of mass loss are for stars at the end point of stellar evolution. Even very massive stars might find means of getting rid of most of their mass and so could end up as neutron stars or white dwarfs.

2. The Gryphon and the Mock Turtle appear in chapters 9 and 10 of *Alice's Adventures in Wonderland.*

3. The Gryphon and the Mock Turtle explain the relation of optical astronomy to the new astronomies which have developed over the last 30 years. Discoveries made in these new astronomical disciplines have led to a more profound understanding of the nature of our Universe. Very often, however, the new astronomies on their own cannot establish the nature of the objects they have discovered but depend heavily upon optical observations to elucidate their nature and physical properties. For example, optical observations often provide the only way of determining the distances of the objects discovered in these wavebands. There is, therefore, a very strong interaction between the new astronomical disciplines and optical astronomy.

4. Alice's statement about the Space Telescope having the greatest gain in sensitivity for starlike objects is absolutely correct. This is because in the case of pointlike objects, most of the light falls on a very small number of the detecting elements (or pixels) of the instruments. For extended objects, the light will be distributed over a larger number of pixels and so there will be a larger contribution from the various sources of background noise present in each pixel.

5. Many of the points Alice makes have already been discussed in note [1]. Here are a few further points.

a. The figure on page 122 shows the luminosity-temperature diagram for a globular cluster. Much of the scatter of the points about the mean line is due to observational uncertainties. With the Hubble Space Telescope, these errors can be greatly reduced and the diagrams extended to much fainter stars.

b. and *c.* See discussion in note [1].

d. With its very high angular resolution, the Hubble Space Telescope will be able to measure the angular distances between stars with a very high degree of precision. An astrometry team led by Dr. William H. Jefferys has been associated with the Space Telescope project. They have demonstrated how the Space Telescope will be able to measure the angular distances between stars with an accuracy of better than $2/1000$ arc second. The ability to measure very accurate parallaxes and other astrometric quantities means that real distances can be measured for objects typically about 10 times more distant than the best that can be done from the surface of the Earth. These measurements will provide much more secure values for many of the fundamental parameters of stars and for the distance scale in our Galaxy.

e. One of the important questions the Space Telescope will address concerns the faintest stars that exist. Theoretically, the centers of stars less massive than about $1/20$ the mass of the Sun do not become hot enough to burn hydrogen into helium. It would be of great interest to know whether or not this limit is actually correct. It is believed that these very faint stars contribute very little to the total local mass density in the Galaxy, but it will be important to check this.

f. There are at least two ways in which the Space Telescope can be used to search for planetary systems around nearby stars. It can search for Jupiter-like objects close to nearby stars using the very high angular resolution capability of the Faint-Object Camera. A simulation of what a Jupiter-like object would look like close to a nearby bright star is shown on page 32. Second, it can look for motions of the barycenter (center of gravity) of the parent star of nearby planetary systems as reflected in small deviations of the motion of the bright star about its mean path across the sky. It was thought that this technique had already demonstrated the existence of planets around several nearby stars, but it is now clear that the deviations seen are not real. The Space Telescope will provide a real opportunity for testing this method. The detection of planetary systems about nearby stars would be one of the greatest discoveries the Hubble Space Telescope could make.

5. *The Violent Interstellar Medium*

1. The Cheshire Cat first appears in chapter 6 of *Alice's Adventures in Wonderland.*

2. The program for the Jargon Phrase Generator is given in *The Computer Book* (London: BBC Publications, 1982), pp. 125–27.

3. THE VIOLENT INTERSTELLAR MEDIUM

One of the areas of the astrophysical research which has developed very rapidly over recent years is the study of the medium between the stars—the interstellar medium. Originally, the interstellar medium was thought to be a rather simple, quiescent medium, but it is now clear from a wide range of different types of observation that it consists of many different phases and components. Its four main constituents are: gas in all its phases, that is, atomic, molecular, and ionized gas; very high energy particles; dust; and magnetic fields. Thus, the interstellar medium is rather complex. It is not stationary but is constantly stirred up by winds blowing from stars, by stellar explosions, and by large-scale perturbations such as the influence of spiral arms.

This new understanding has come from observations in the many different astronomical wavebands which have been opened up over the last 20 years. All bodies emit radiation, and the characteristic wavelength at which the radiation is emitted is directly related to the temperature of the emitting body. Thus, bodies with temperatures of about 3000 degrees Kelvin typically emit radiation in the optical waveband. All temperatures are quoted in degrees Kelvin (K), which is approximately the temperature in degrees Celsius (C) plus 273. Thus, a warm day at 27 degrees Celsius corresponds to 300 degrees Kelvin, or 300 K. This temperature scale in degrees Kelvin is measured from the absolute zero of temperature at approximately $-273°$ C. Bodies at room temperature, which is about 300 K, emit in the infrared waveband at wavelengths ten times longer than at optical wavelengths, and cool materials with temperatures in the range 3 to 30 K ($-270°$ to $-243°$ C) emit in the millimeter and submillimeter wavebands. In the same way very hot gasses emit at wavelengths shorter than the optical waveband. Gas at 30,000 K emits strongly in the ultraviolet waveband, and very hot gasses, at temperatures of a million degrees or greater, are strong X-ray emitters. This relationship between temperature and wavelength is shown on pages 128 and 129.

With the advent of different types of telescopes which can make observations in these wavebands, it has become apparent that gases at a very wide range of temperatures and densities are present in the interstellar medium. The main features of the different components of this medium can be summarized as follows.

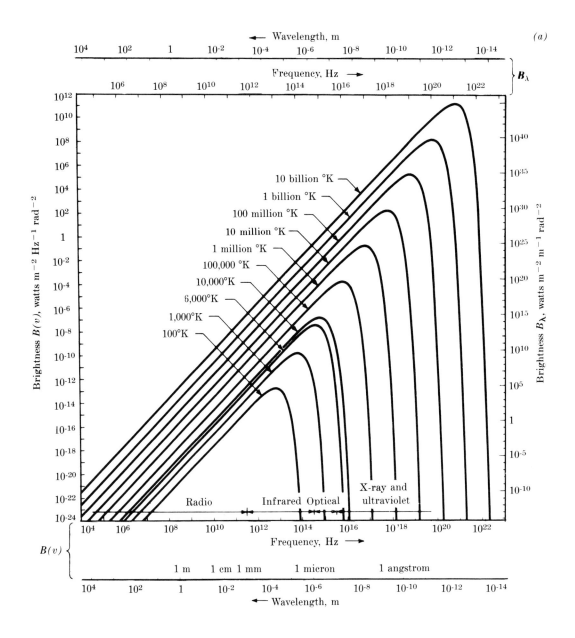

(a) The spectrum of black bodies at different temperatures. The higher the temperature, the greater the total amount of radiation emitted by the body and the higher the frequency (or shorter the wavelength) at which most of the radiation is emitted. *(b)* The relation between the temperature of a black body and the frequency (or wavelength) at which the maximum of the radiation spectrum occurs. The different wavebands are indicated on each diagram.

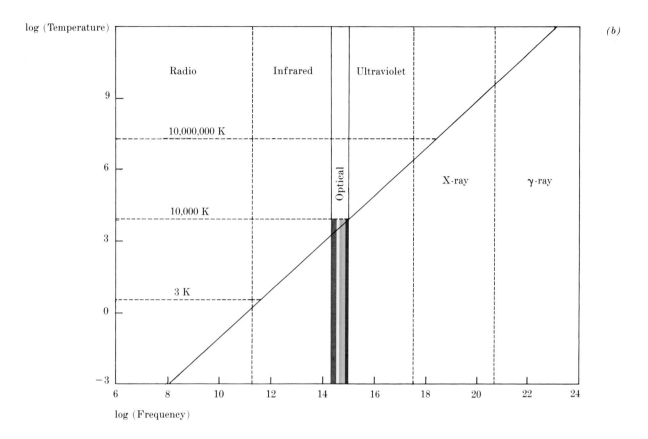

(b)

Gas

The coolest components observed are the giant molecular clouds, which are present throughout the interstellar medium. These are observed by radio telescopes working at millimeter and submillimeter wavelengths through their intense molecular line emission. Many molecular species have been detected in the interstellar gas in giant molecular clouds, and Alice gives a recent compilation of the molecules detected so far. A number of these have only been detected in the interstellar medium. Evidently, there is a great deal of organic and inorganic chemistry going on in the interstellar gas. Typical giant molecular clouds have mass about a million times the mass of the Sun and their temperatures lie in the range 10 to 100 K. Molecular line and infrared observations have shown that much denser regions exist within the large clouds and it is within these that stars are formed. Hot molecular gas is observed close to the regions where stars have already formed. The molecular gas is heated by the radiation of the newly formed stars as it pushes back the surrounding medium.

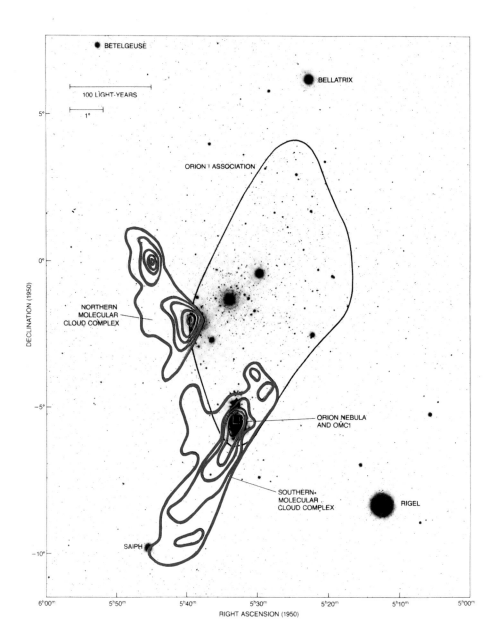

A superposition of the distribution of molecular radio line-
radiation of carbon monoxide (red contours) upon an optical
image of the constellation of Orion. The radio emission comes
from a giant molecular cloud, which is very much larger than
the famous Orion Nebula, the small emission-line region halfway
down Orion's sword. The black contour outlines a region in which
many bright stars are found in a large cluster, known as the
Orion 1 Association.

[130]

Hot ionized gas at temperatures about 10,000 K is observed around young stars. The gas is sufficiently hot to remove the outer electrons of the atoms. The beautiful diffuse structures seen in objects like the Orion Nebula or the Horsehead Nebula (see pp. 37 and 38) are due to the strong emission lines of this ionized gas. The hot young stars are exciting the surrounding region and causing it to radiate.

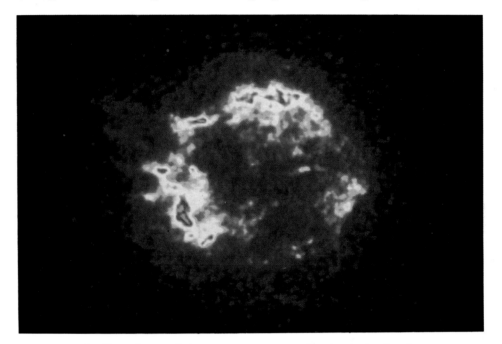

An X-ray image of the supernova remnant Cassiopeia A taken by the Einstein X-ray Observatory. The X-rays are the emission of very hot gas expelled in the explosion of the star and also interstellar gas heated up by the expanding shell of the remnant.

In addition to these components associated with the regions where stars are formed, there is very hot gas which is ejected in supernova explosions. These very violent events in which the whole star is disrupted liberate a huge amount of energy which is sufficient to heat up the gas in a large volume around the explosion to very high temperatures. The overlapping of the outflowing material from the explosions of different generations of stars results in a considerable fraction of the interstellar gas being heated to these high temperatures. This gas is detected through its X-ray emission and through the observation of absorption lines in the interstellar gas due to highly ionized species. Its temperature is typically about a million degrees Kelvin.

There is, in addition, between the very hot regions and the giant molecular clouds, diffuse gas which consists of a mixture of ionized and un-ionized gasses. As has been emphasized, the different components are not stationary but are shocked by supernova explosions and by strong stellar winds. These motions compress the gas and may assist in the formation of molecular clouds.

[131]

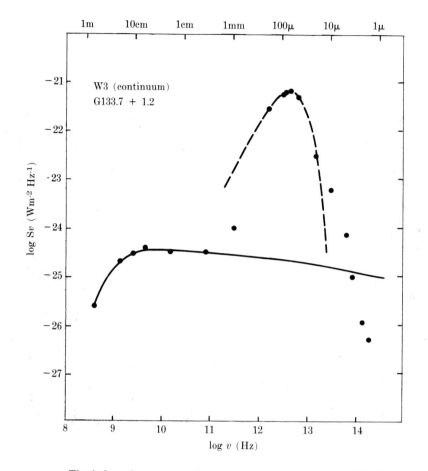

The infrared spectrum of the region of star formation W3. Most of the energy radiated by W3 is emitted in the far-infrared region of the spectrum and is the radiation of dust heated by very young stars or protostars.

Interstellar Dust

The black areas seen in photographs such as those of the Orion and Horsehead Nebulae are due to obscuring dust. Dust is present throughout the interstellar gas and has the unfortunate effect of obscuring some of the most interesting regions we wish to study. Fortunately, it becomes transparent to infrared radiation, thereby making it possible for us to see deep inside the regions where stars are forming. In addition, the dust is heated by the radiation it absorbs and is therefore a strong emitter at the temperatures to which it is heated, which are typically in the range from about 30 to 500 K. Indeed, in the infrared waveband the very dusty regions become the most intense emitters while they remain the most obscuring objects when observed in the optical waveband. There is a good reason why the dust radiates at temperatures less than about 1000 K, and that is because at higher·temperatures, it would evaporate and so would no longer exist as dust. Most of the dust particles have dimensions of about 1 micron (one-thousandth of a millimeter) and are similar in size to particles of cigarette smoke, which explains why the Cheshire Cat's room is so smoky. The dust plays a crucial role in the processes by which stars are formed. First, it protects the fragile molecules from being dissociated by the intense interstellar radiation field, and second, it acts as an efficient means by which the protostar can lose energy by radiation. Indeed, objects which are likely to be protostars in the regions where stars are formed are characterized by very intense emission in the infrared waveband (see p. 49).

Interstellar High-Energy Particles

Very high energy particles are present throughout the interstellar medium. These particles are accelerated in supernova explosions and are then dispersed throughout the interstellar medium. The electron component is detected by its radio emission,

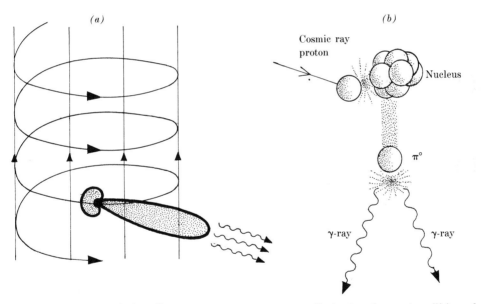

Synchrotron radiation of
high energy electrons

Production of γ-rays by collisions of
cosmic rays with interstellar nuclei

Diagrams illustrating the processes by which high-energy particles can be detected astronomically. *(a)* High-energy electrons emit highly beamed radiation when they spiral in a magnetic field. This form of radiation is called synchrotron radiation. *(b)* When high-energy protons or nuclei collide with the nuclei of atoms and ions of the interstellar gas, pions are generated. The neutral pions produced decay very rapidly into two gamma rays, which can be detected by gamma ray telescopes.

which is due to a process known as sychrotron radiation. In this process, the very high energy electrons emit radio waves as they spiral in the weak magnetic field present in the interstellar gas. High-energy protons are also observed through the gamma-rays which they emit when they collide with ordinary matter in the interstellar gas. These very high energy particles are very rare, but they are so energetic that they have an important effect upon the dynamics of the interstellar gas.

The Galactic Magnetic Field

Finally, a number of separate observations indicate that there is a weak, large-scale magnetic field present in the interstellar gas. The evidence comes from radio observations of the emission of high energy electrons in our Galaxy and from the polarization of the radio emission and of the light of nearby stars. This magnetic field has an important influence on the dynamics of the different components of the interstellar gas (see note [5]).

4. STAR FORMATION

The Cheshire Cat is not alone in not understanding the precise details of how stars are formed. This is probably the most important unsolved problem of the astrophysics of stars and galaxies. We do not know how many massive and how many light stars are formed in any particular molecular cloud. We do not know how the rate at which stars form depends upon physical conditions within the cloud. We do not know how the molecular clouds themselves form. Some of these problems can be addressed by the Hubble Space Telescope and others by observations in the new disciplines of infrared and millimeter astronomy. The millimeter wavebands are at a relatively early stage of exploitation for the study of these problems, and very significant progress will undoubtedly be made in these areas during the next few years as new and powerful telescopes and instruments are brought into use.

The theoretical position is not very much better. The fundamental problem to be understood is the sequence of events which must take place between the formation of the molecular cloud, which has a typical density of about 1000 molecules per cubic centimeter, and the formation of a new star, which has a density billions and billions of times greater. The denser regions of molecular clouds will collapse under their own gravity, but as they collapse they heat up and the collapse slows down. There must be some means by which the heat generated in the collapse can be removed from the cloud. Radiation is almost certainly the answer. However, because the radiation is trapped inside the collapsing cloud, the interior heats up. The slower collapse and heating up continues until the center of the protostar becomes so hot that nuclear burning of hydrogen into helium begins and the star starts its life as a main-sequence star. Although this seems a very straightforward process, it presents many grave theoretical uncertainties. For example, it is not clear how one can get rid of any rotation or magnetic fields in the collapsing cloud. These are strongly amplified during collapse and can prevent collapse unless some means is discovered of removing the rotation and magnetic fields from the region.

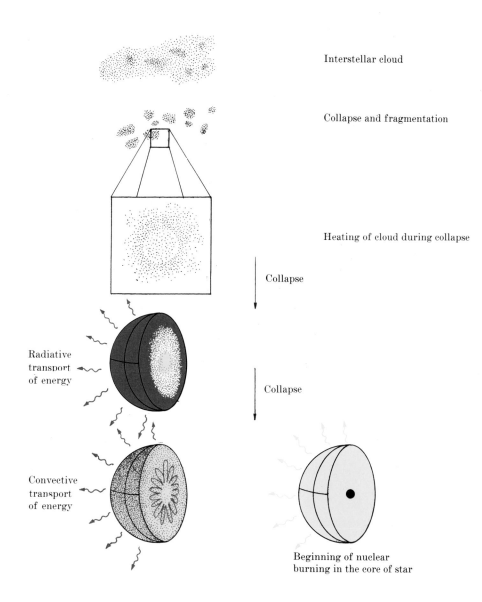

Interstellar cloud

Collapse and fragmentation

Heating of cloud during collapse

Collapse

Radiative transport of energy

Collapse

Convective transport of energy

Beginning of nuclear burning in the core of star

A schematic diagram showing a possible sequence of events which leads to the formation of stars. A giant molecular cloud fragments into smaller and smaller cloudlets, and these fragments collapse to form new stars. Initially the collapsing star can lose energy directly from its interior by radiation. However, when the cloud becomes opaque to radiation, the energy takes much longer to escape. The transport of energy to the surface is speeded up when the interior becomes unstable and large convection currents are set up. As the center heats up, it eventually reaches a temperature at which nuclear reactions can take place, and the star begins its life as a main sequence star.

These studies have implications far beyond their immediate relevance for the formation of stars. For example, in order to understand the evolution of galaxies, we have to know the rate at which new stars are forming under a wide range of different conditions. This field of study is certainly one of major importance for essentially all branches of contemporary astronomy and cosmology.

5. The rubber bands which the Cheshire Cat and his students attach to everything in their room are intended to represent the magnetic field in the interstellar gas. It turns out that in many ways magnetic fields behave like rubber bands in these conditions. The presence of even a very small amount of ionized gas is sufficient to tie the matter to the magnetic field; when the gas moves the magnetic field moves with it. This phenomenon is known as *magnetic flux freezing* because the magnetic field is effectively frozen into the gas. If the gas is compressed, the strength of the magnetic field increases. If the field is stretched, the tension in the magnetic field increases, and so on. From this picture, it can be understood why the magnetic field can impede the collapse of a gas cloud: because the "rubber bands" remain attached to the rest of the interstellar medium, there must be some means of ridding the collapsing region of magnetic fields.

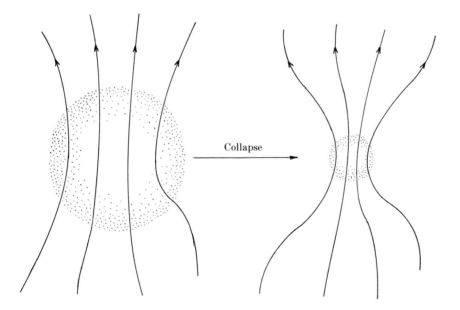

Illustration of the process of magnetic flux freezing. In an ionized plasma the magnetic field lines are tied to the plasma, and as the plasma moves, the magnetic field lines move with it. Here the collapse of the cloud results in the amplification of the strength of the magnetic field.

6. *The Building Blocks of the Universe*

1. DISTANCES IN THE UNIVERSE

One of the most common questions which I am asked by nonastronomers is how one can envisage the vast distances we have to deal with in astronomy. The answer is simple. One never thinks in terms of absolute numbers, such as the number of centimeters to the nearest star or the edge of the Universe, but rather in relative terms, so that one asks how far away one object is relative to the size or distance of another. Since we are about to embark on the first stages of a journey to the edge of the Universe, let us show how we can arrive at these large scales in six easy stages from dimensions with which we are familiar within the Solar System.

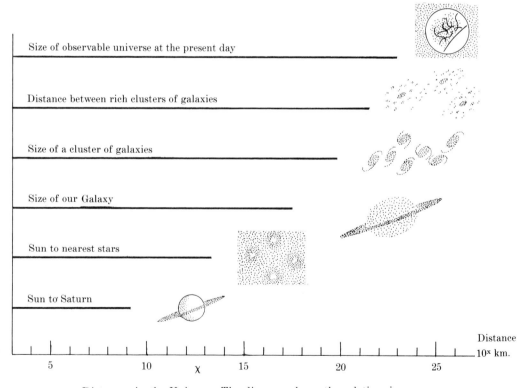

Distances in the Universe. The diagram shows the relative sizes or distances of objects in logarithmic terms. To put it simply, the numbers on the horizontal axis show the numbers of zeros which should be put after 1 to obtain the distance in kilometers.

[137]

A photograph of the Pleiades, one of the star clusters closest to the Earth. The cluster is only about 20 million years old and lies at a distance of approximately 400 light-years.

We are all familiar with the distance between the Sun and the Earth. It is about 150 million kilometers (or 93 million miles). Light travels at a speed of about 300,000 kilometers (or 186,000 miles) per second, and therefore light takes about eight minutes to travel from the Sun to the Earth. It is convenient to measure these large distances in terms of the time it takes light to travel that distance, and so we say that the Sun is at a distance of about eight light-minutes from the Earth.

Distances within the Solar System are now familiar to us; space probes have already been sent to Venus, the Moon, and Mars, and spacecraft have flown past the more distant planets—Mercury, which is the closest to the Sun, and the giant outer planets, Jupiter and Saturn. The distance from the Sun to Saturn is about 10 times the distance from the Sun to the Earth, that is a distance of about 80 light-minutes. Because space probes such as *Voyager 1* and *Voyager 2* travel much slower than the velocity of light, the journey to Saturn from the Earth took about four years.

To reach the nearest stars, we have to travel a distance about 20,000 times greater than the distance from the Sun to Saturn. This corresponds to a distance of about 4

A picture of the whole sky painted by M. and T. Keskula of the
Lund Observatory. The Milky Way runs along the center of the
diagram, and the center of the Galaxy is in the center of the
picture. The coordinates are squashed so that equal areas are
preserved. This picture shows that our Galaxy has the shape of a
flattened disk.

light-years. Within a distance of 17 light-years, we know of about 65 stars. Several of
them are similar to our own Sun, but most are fainter and only four are brighter.

The stars we see nearby are only a few of the 200 to 400 billion stars which make up
our own Galaxy. The Milky Way, which can be seen clearly on a dark night as a hazy
band across the sky, is the integrated light from all these stars . The Milky Way forms
what eighteenth- and nineteenth-century astronomers referred to as an Island
Universe ; that is, the stars we see locally are not distributed uniformly throughout
the Universe but are clustered into vast stellar associations known as galaxies. If we
could look at our own Galaxy from outside, it would probably resemble our giant
companion in space, the Andromeda Galaxy, also known as M31 (see p. 45). Our
Galaxy is disk-shaped, with a pronounced central bulge. The Solar System is located
towards the edge of the disk, on the inner edge of a spiral arm similar to those seen in
other spiral galaxies. The distance from the Solar System to the center of the Galaxy
is about 30,000 light-years, or roughly 7,000 times the distance from the Sun to the
nearest stars. It is not very meaningful to give a single number which corresponds to

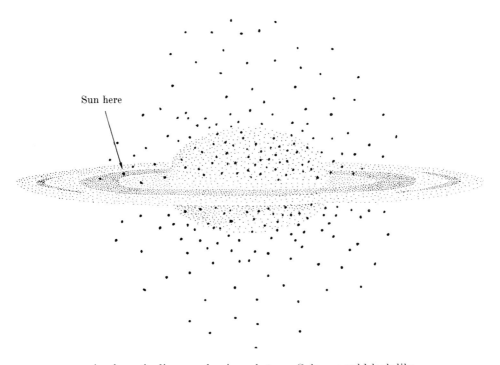

Sun here

A schematic diagram showing what our Galaxy would look like
from outside. The disk component contains the youngest stars in
the Galaxy, the youngest of all being concentrated in spiral
features. The "bulge," or spheroidal component, contains the
oldest stars in the Galaxy. Among these are the globular clusters
which are part of the "halo" population.

the total diameter of the Galaxy, because it does not have a definite edge. The visible
image of a galaxy such as our own has a size of about 100,000 light-years.

The galaxies define the large-scale structure of the Universe. Galaxies have a
strong tendency to be clustered. This clustering ranges from pairs and loose groups
to giant rich clusters, which are among the most striking objects in the sky. The Pavo
cluster, which contains well over 1000 galaxies, is shown on p. 50. The typical size of a
cluster of galaxies is about 100 times the size of our own Galaxy. Rich clusters of
galaxies such as the Pavo cluster are quite rare and are far outnumbered by smaller
groups and associations, which contain most of the galaxies in the Universe.

Going yet farther up in scale, what is the typical distance between these rich
clusters of galaxies? The separation between the rich clusters is about 20 to 30 times
their typical size.

Finally, we can ask how much bigger the Universe is than the typical distance
between rich clusters of galaxies. The best way of giving an unambiguous answer is
to ask, How far away can we observe the Universe more or less as it is today? The
problem is that when we look very far away, we are looking back in time, because light
travels at a finite speed. Therefore, the best estimate of the size of the observable

Universe more or less as it is now corresponds to looking back to when the Universe was about half its present age. This distance is roughly 50 times the separation between rich clusters of galaxies, or about 15 billion light-years. It becomes less and less meaningful to talk about distances greater than this because we are then looking so far back in time that the Universe was much younger than it is now and time becomes a much more useful measure of the Universe than distance. We will go into this aspect of our story in greater detail in the notes accompanying chapter 8.

Notice how we have come from the scale of the Solar System to the size of the observable Universe in six easy stages, at no stage going up by a factor of more than 20,000 in size. All this means is that it is much easier to think in relative rather than absolute terms in astronomy and cosmology. To put it another way, it is easiest to think logarithmically about distances in astronomy.

These distances are summarized in Table 1.

We have used light-years as our unit of distance in the above discussion: 1 light year is 9.46×10^{12} km. Astronomers normally use a different unit of distance, known as a *parsec*, or *parallax-second* (abbreviated to *pc*), which corresponds to 3.086×10^{13} km. One parsec is roughly 3.26 light-years. The parsec is defined as the distance at which the semimajor axis of the Earth's orbit about the Sun subtends an angle of 1 arc second. One megaparsec (Mpc) = 1 million parsecs = 3.086×10^{19} km.

Table 1   Physical Sizes and Distances in the Universe

	Distance or Size	Relative Size
1. Solar System Sun to Saturn	1,400,000,000 km = 1.4×10^9 km	10 times the distance from Sun to Earth
2. Nearest stars Sun to nearest stars	3×10^{13} km	20,000 times the distance from Sun to Saturn
3. Our Galaxy	10^{18} km	30,000 times the distance from Sun to nearest stars
4. A cluster of galaxies	about 10^{20} km	100 times the size of a galaxy such as our own
5. Distance between rich giant clusters of galaxies	about 3×10^{21} km	20–30 times the size of a cluster of galaxies
6. The observable Universe at the present time	about 1.5×10^{23} km	about 50 times the distance between rich clusters

2. Galaxies are the basic building blocks of the Universe. Most of their mass is in the form of stars. It is the gravitational pull of the stars on one another which holds a galaxy together. (See, however, note [7], concerning the problem of the dark or hidden mass which may be present in the outer regions of giant galaxies.)

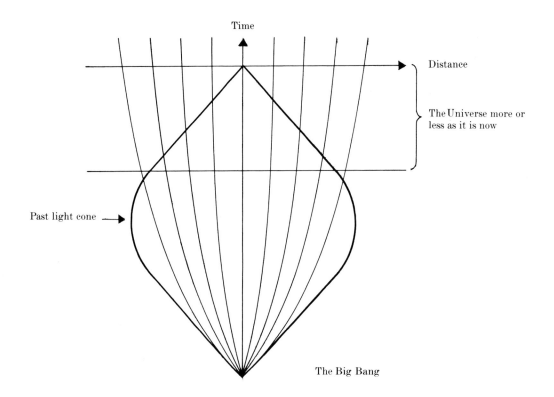

A simple space-time diagram illustrating what we mean by "the Universe at the present day." Time runs up the diagram and distance across it. When we observe distant objects, we look into the past. Eventually we look so far away that the Universe must have been very much younger when the light was emitted by the distant object than it is today.

Many different types of galaxy have been identified, but the basic distinction which is obvious on a photograph is between spiral and elliptical galaxies. A smaller fraction of galaxies are known as irregular galaxies.

The spiral galaxies have a disk shape with a central bulge like that of our own Galaxy. The relative sizes of the disk and bulge vary from galaxy to galaxy. Their masses range from systems from 10 to 100 times more massive than our own Galaxy, which has mass about 100 billion times the mass of the Sun, to dwarf systems, which are only about 10 million times the mass of the Sun. The disks of spiral galaxies rotate and this is what gives them their characteristic shapes. There are normal spirals (see, e.g., p. 45) and also barred spiral galaxies (see p. 45) in which the central bulge is elongated and the spiral arms trail from the ends of this "bar." The spiral arms are defined by the youngest (hottest, brightest) types of star and also by the ingredients of regions of star formation, that is, by gas clouds and dust. There is continuing star formation in the arms of spiral galaxies.

The elliptical galaxies have much smoother profiles, with little evidence of dust, gas, or spiral arms in general (see p. 46). They are spheroidal in shape and can have masses from about 100 times the mass of our own Galaxy to only about 10 million times the mass of the Sun. They are self-gravitating systems in which the random velocities of the stars prevent the galaxy's collapse.

There are galaxies which appear to be intermediate between the spirals and the ellipticals which are known as SO, or lenticular, galaxies (see p. 51). They look rather like spiral galaxies which have been stripped of their spiral structure; that is, they possess stars in a disk and have a central bulge but have little evidence of spiral arms.

The irregular galaxies (see p. 47) are generally less massive than typical spiral and elliptical galaxies. They have an irregular structure and often have large amounts of gas and dust. In a number of cases it is very likely that the irregular appearance is due to the galaxy having had a strong gravitational interaction with another nearby galaxy.

There are in addition a number of special classes of galaxy. Perhaps most interesting astrophysically are the galaxies with active galactic nuclei (see Chap. 7). The most energetic of these active nuclei are rather rare phenomena, but the centers of most galaxies probably possess small-scale versions of the activity seen in some of the most active galaxies and quasars.

3. This simulation of what may be seen in a deep-sky image taken with the Wide Field/ Planetary Camera was kindly provided by Dr. John Bahcall. The image corresponds to an area on the sky of 1.33 × 1.33 arc minutes square, and Bahcall and Soneira predict that in this area there will be 51 galaxies and 6 stars. The typical redshift of the galaxies is probably about 1, meaning that the galaxies emitted the light we see when the Universe was about half its present age.

4. The Mad Hatter, the March Hare, and the Dormouse have their tea party in chapter 7 of *Alice's Adventures in Wonderland*.

5. One of the aims of the astrophysics of galaxies is to understand why there is such a very wide diversity in their properties. The aim of astrophysicists is to put some order into this diversity. The mass is one of the most important parameters of galaxies because so many properties are correlated with the mass or luminosity of the galaxy. Whenever a correlation of this sort is discovered, it provides important clues about what it is that determines the similarities and differences between galaxies. The angular momentum is also likely to be important, as are features like the environment in which the galaxy finds itself. For example, does the galaxy lie in a cluster or is it isolated?

A simulation by Drs. J. Bahcall and R. Soneira of what the sky
in the direction of the North Galactic Pole may look like when
observed in a long exposure with the Wide Field/Planetary
Camera. The faintest stars visible are about 28 magnitude. The
area of the photograph is only 1.33 × 1.33 arcmin square. The
actual pictures taken with the Wide Field/Planetary Camera will
cover four times this area.

6. In many respects, the stars of a galaxy may be thought of as resembling the atoms or molecules of a gas. The big difference between them is that in a typical gas the atoms collide very frequently, thereby enabling energy to be exchanged between atoms until eventually the particles of the gas all have the same average energies. The stars in a galaxy behave very differently. They do not collide with one another, and as a result they do not exchange energy but rather preserve some information about their initial dynamical properties. There is now convincing evidence that satisfactory models of elliptical galaxies must take into account the fact that the stars do not have the simple distributions found in a gas. For example, it has now been demonstrated by observation that the flattening of large elliptical galaxies cannot be attributed to rotation of an otherwise simple galaxy in which the velocities of the stars are randomized. This observation shows that in these galaxies energy has not been shared out equally among all the motions of the stars in a galaxy, and unfortunately many of the simple approximations which can be made in the study of gases can no longer be used. Considerable advances in these studies have been made in recent years through the work of Professor Martin Schwarzschild at Princeton University Observatory concerning techniques for building theoretical models of galaxies in which the particles have essentially no interaction with one another but only with the gravitational field of the galaxy as a whole.

7   THE PROBLEM OF THE DARK, OR HIDDEN, MATTER

In astronomy, all methods of estimating masses are dynamical. For example, to measure the mass of the Sun, we work out the velocities of the planets about the Sun and measure their distances from the Sun. Then, using Newton's laws of motion and of gravity, we can work out the mass of the Sun because it is the attractive force of gravity which keeps the planets in their elliptical orbits. This illustrates the important point that to measure masses in astronomy, we need measures of the velocity and size of the system under consideration.

This procedure can be applied to many systems in astronomy. For example, the masses of stars in a binary system are worked out from their orbital velocities and the distance between the components of the double star. The masses of globular clusters and elliptical galaxies can be found by measuring the size of the systems and the velocities of the stars within them. In the case of spiral galaxies, the rotational velocities in the disk and its variation with distance in the galaxy enable the mass and mass distribution to be worked out. Finally, in clusters of galaxies, the total mass can be worked out by measuring the size of the system and the dispersion in velocity of the galaxies about the mean velocity of the cluster.

The problem of the dark, or hidden, matter arises as follows. The masses of galaxies and clusters of galaxies can be measured by the above procedures. It is found that in the giant clusters, there must be about 10 times more mass present in the cluster than can be accounted for by the galaxies alone. To account for the total mass of the cluster, there must be this unseen mass, which so far has only been detected by its gravitational influence—this mass is referred to as dark, or hidden, matter.

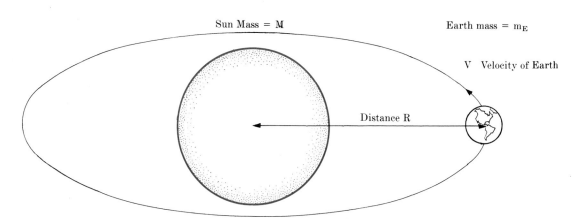

Sun Mass = M   Earth mass = m_E

V   Velocity of Earth

Distance R

Gravitational attraction of Sun = Centrifugal force of Earth's motion about the Sun

$$\frac{GMm_E}{R^2} = \frac{m_E V^2}{R}$$

$$M = \frac{V^2 R}{G}$$

Illustration of how astronomers measure the masses of celestial
objects. All direct measurements of mass in astronomy use
Newton's law of gravitation. Here the mass of the Sun is found
by measuring the Earth's velocity about the Sun and the radial
distance of the Earth from the Sun.

The problem is now believed to exist not only for giant clusters of galaxies but also for groups of galaxies and in the outer regions of giant spiral and elliptical galaxies. The nature of the dark matter has not been established and the Dormouse gives a list of some of the possibilities. The point is that there are many forms of matter which could be present in the Universe and which would be very difficult to detect observationally. The possibilities range from neutrinos and other very weakly interacting particles to massive black holes. There is no agreement about the probable nature of the dark matter, but it is plainly a key problem which may have implications for fundamental physics.

8.   I apologize for perpetrating another Carroll parody, this time on "The Walrus and the Carpenter" from *Through the Looking-glass*, chapter 4. Professor Jeremiah Ostriker is Director of the Department of Astrophysical Sciences, Peyton Hall, Princeton University. Professor Ostriker and his student Mark Hausman wrote two fundamental papers about the problem of galactic cannibalism. By this we mean that one galaxy can consume another, resulting in a system which is more massive than either of the originals. This process is particularly important in rich clusters, where the distances between galaxies is small and encounters between galaxies are there-

fore rather common. My poem describes what Ostriker and Hausman found would be likely to happen in clusters of galaxies. Some of their most important results came from computer simulations of how the process of cannibalism would take place in a realistic model for a cluster.

The likely sequence of events can be described as follows. First, when the cluster forms, the galaxies or protogalaxies will not have the spatial distributions which we now observe in clusters. The galaxies will collapse towards the center of the cluster and then the process known as *violent relaxation* enables the galaxies in the cluster to settle down to the roughly symmetric distribution which we observe in clusters now. In the process of violent relaxation, first described by Professor Donald Lynden-Bell, the galaxies are redistributed by large irregularities in the gravitational field of the collapsing cluster. After this stage, the galaxies begin to evolve towards an equilibrium distribution. In this process, the most massive galaxies are slowed down and therefore fall towards the center of the system, whereas the lighter ones are speeded up. The decelerating effect due to the gravitational influence of individual galaxies in the cluster is called *dynamical friction*. The most massive galaxies thus drift quickly into the center and then begin to disrupt and consume less massive galaxies which come close to the center. In this way, the central galaxy can become very large and bloated at the expense of the less massive galaxies. Theoretically, it can be shown that the most massive galaxies are eaten first and then the lighter ones. This appears to be what is found in rich clusters of galaxies in which the more luminous the central galaxy, the less luminous are the next brightest members.

An understanding of this process is important for cosmology because the most massive galaxies in clusters are among the best tools for studying the Universe at large distances and for measuring cosmological parameters like the deceleration parameter (see Chap. 8).

The reader will understand that a certain amount of poetic license has been taken to make the characters of the present poem match the originals, whose thoughts are dominated by eating. To my knowledge, neither Dr. Ostriker nor Dr. Hausman is afflicted with this obsession with eating.

9. Professor James Gunn is one of the astronomers who has made major contributions to the study of the evolution of galaxies and of cannibalism by giant galaxies in the centers of rich clusters of galaxies.

10. The question of interactions between galaxies is becoming very important for a variety of different problems in astrophysics. The possibility of feeding black holes will be described in chapter 7. When spiral galaxies collide, the gas and dust are strongly heated, and it may well be that it is the radiation from this dust which has been observed from interacting galaxies by the *IRAS* satellite.

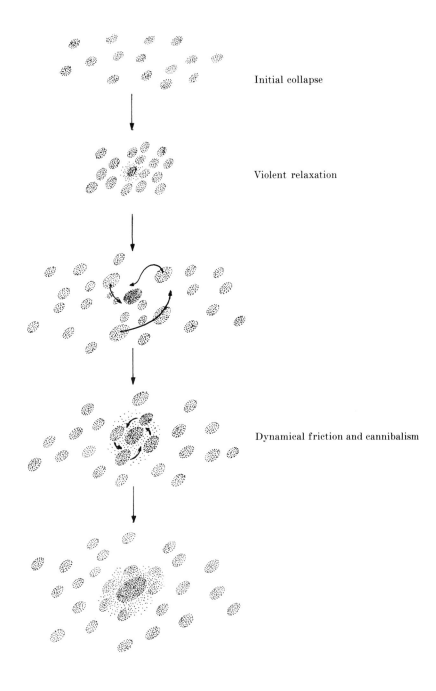

Initial collapse

Violent relaxation

Dynamical friction and cannibalism

A schematic diagram showing a possible picture of the evolution of a cluster of galaxies. Shortly after the collapse of the cluster, the galaxies take up a more, or less regular pattern; then the most massive members spiral into the middle and a supergiant galaxy forms at the center by the coalescence of these galaxies.

7. *Active Galactic Nuclei and Black Holes*

1. American readers will recognize the warning which appears on all cigarette advertisements.

2. Alice meets the Caterpillar in chapters 4 and 5 of *Alice's Adventures in Wonderland*.

3. ACTIVE GALACTIC NUCLEI AND QUASARS

Active galactic nuclei comprise one of the most important fields of current research in extragalactic astronomy. We know of many "active" systems in astronomy—for example, supernovae, in which stars explode, and pulsars, which emit strong radio emission. We also know that optical photographs of galaxies often suggest that there is a great deal of activity in some galaxies. However, the terms *active galaxy* and *active galactic nucleus* mean something quite specific; they refer to those galaxies in which there is direct observational evidence for a very compact and highly energetic source in which the energy is not produced by the normal processes we associate with stars.

Although there are many separate pieces of evidence which lead to this picture, Alice summarizes the one key feature which has to be explained. In certain galaxies there is evidence for enormous amounts of energy being liberated over very short time scales indeed. NGC 4151 is a good example of this phenomenon—its nucleus can suddenly liberate a huge amount of energy in a few days or weeks. During the outburst, the luminosity of the nucleus may be billions or trillions times greater than that of the Sun.

It is useful to give a catalogue of the various types of active nuclei which are now known. Historically, the *Seyfert galaxies* were the first class of galaxies with active nuclei to be discovered. Like NGC 4151, they appear to be spiral galaxies, but they possess starlike nuclei (see p. 62). In addition, when the spectra of these galaxies are studied, the emission lines are found to be very broad and very strong, unlike those found in hot regions of ionized gas in our Galaxy. The Seyfert galaxies are among the most important classes of active nuclei because they are relatively common and are found in reasonably nearby systems, thus enabling their properties to be studied in detail.

The next objects to be discovered were the *radio galaxies,* so called because the first radio surveys showed them to be intense emitters of radio waves. By the mid 1950s it was established that these galaxies must be sources of vast amounts of high energy particles and magnetic fields. A few of these galaxies, which were called *N galaxies,* had starlike nuclei similar to those of the Seyfert galaxies and also had strong broad emission lines in their spectra, but the relation between these phenomena was not clear.

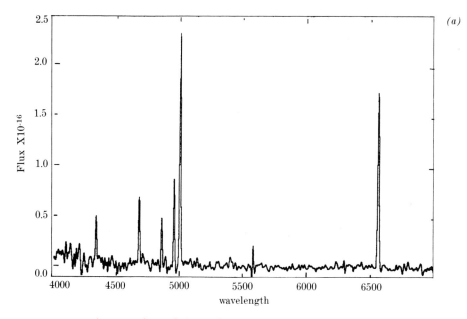

A comparison of the optical spectra of *(a)*, a region of hot ionized hydrogen in our Galaxy associated with the planetary nebula NGC 246, and *(b)*, the nucleus of the active galaxy Markarian 335. There are two important differences between the spectra. First, the lines in the active nucleus are very much broader than those of the hot gas cloud, the widths of the lines corresponding to velocities of about 1000 kms[-1]. Second, there are different ionic species present in the two cases. In the active nucleus, a very much wider range of excitation conditions is observed than in the planetary nebula.

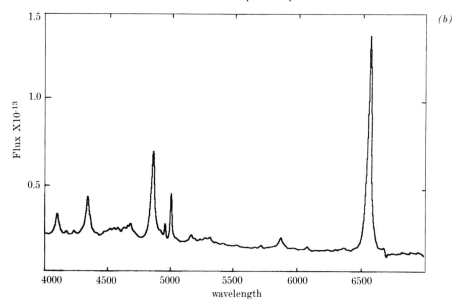

Then in 1962, Maarten Schmidt measured the first redshift of a *quasi-stellar radio source*, or *quasar*. The radio source 3C 273 had been discovered in early radio surveys of the sky. The remarkable thing about 3C 273 was that the optical counterpart of the radio source looked exactly like a star on a photographic plate and was found to be variable. Yet it turned out to lie at a distance similar to that of the most distant galaxies whose distances were readily measurable at that time. The object was called a quasi-stellar object because it looked like a star but clearly it could not be any normal sort of star at this very great distance. The optical luminosity of 3C 273 is more than a thousand times greater than that of a galaxy such as our own. Following this great discovery, many more quasars were found, all of them characterized by stellar appearance, very great distances, and vast luminosities. In fact, quasars are the most distant objects we can observe in the Universe, the most distant of them having emitted their light when the Universe was about one-fifth its present age.

Quasars are among the most extreme examples of active galactic nuclei known. We know that they actually are the nuclei of galaxies because it is possible to observe the underlying galaxy in some of the closer examples.

Perhaps the most extreme examples of active galactic nuclei are the objects known as *BL-Lacertae* (or *BL-Lac*) *objects*. They are similar to the quasars but differ from

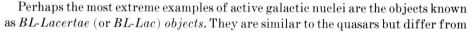

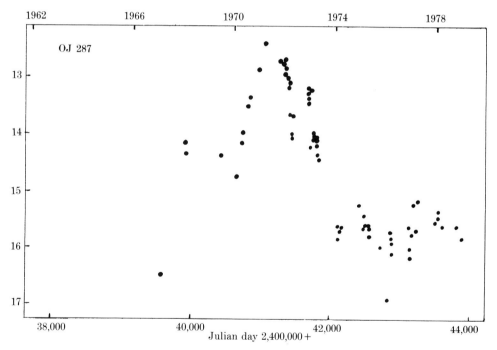

The optical variability of the BL-Lac object OJ 287 is illustrated
by its light curve, which spans the years 1966 to 1979. Its
luminosity is quoted in magnitudes so that 2.5 units on the
vertical axis corresponds to a factor of 10 in intensity. OJ 287
has varied in intensity by a factor of almost 100 throughout this
period. In addition, very rapid fluctuations in intensity occur on
the time scale of days or less.

them in two important respects. First, they vary extremely rapidly in luminosity, variations being detected on time scales of days or less; they must therefore be very compact objects. Second, they normally have no strong emission lines in their spectra. It is plausible that in the BL-Lac objects we observe most directly the primary source of energy in active nuclei.

It should be emphasized that the most active of these systems are very rare. They are relatively easy to discover because they are so luminous and have such distinctive properties. The astrophysical problem is to understand how they can produce such vast amounts of energy from such compact volumes.

The current consensus is that these different types of active nuclei are all basically the same. In these objects, there is some ultracompact source of energy and precisely what one observes depends upon the environment of the compact object and the way in which it obtains its fuel. Even rather quiescent nuclei like the center of our own Galaxy probably possess mini versions of the phenomena we see in quasars and other active nuclei. It is likely that there is a continuity in activity which runs from systems like our own Galaxy through the Seyfert galaxies and N galaxies to the quasars and BL-Lac objects.

Let us summarize the principal observational characteristics of these active galactic nuclei. First, the nucleus appears to be starlike and may exhibit variability over time scales from less than days to years. Second, the light from the active nucleus is not starlight but is due to the radiation of high energy particles, which originate in very large quantities in the nucleus. These particles are also likely to be responsible for the variable X-ray emission often observed from these objects. Third, the optical spectrum is often characterized by the presence of very strong emission lines of the common elements. These lines are much stronger and broader than those seen in normal galaxies, and there is convincing evidence that the line-emitting regions are excited and heated by the optical and ultraviolet radiation generated in the nucleus. A number of arguments suggest that the strong line emission is likely to originate rather close to the nucleus itself, among the most convincing evidence being the variability of the broad emission lines over short time scales, as in the case of NGC 4151. In a few cases, these active galaxies are also strong radio emitters, and we will look at this aspect of their properties in note [10].

4. BLACK HOLES

Black holes have become part of the astronomical landscape very quickly. We have already described the ultimate fate of stars in Chapter A4, note [1]. The basic question is whether or not there exist pressure forces inside the star which can prevent its collapsing under its own gravity. In normal stars, support is provided by the pressure of a normal compressed gas mixed with a great deal of radiation, all the energy being generated in nuclear reactions in the center of the star. In white dwarfs, pressure support is provided by the degeneracy pressure of the electrons. This means that the electron gas is squeezed so hard that quantum mechanical forces keep the gas from being squashed any further. In neutron stars, the degeneracy pressure support is provided by the same quantum mechanical forces, but now associated with neutrons and protons—this is called neutron degeneracy pressure. Neutron stars are very dense indeed, densities of approximately a million billion times that of ordinary matter being found in their interiors. At this stage, the force of gravity is very strong and the radius of the star is only about three times that of the corresponding black hole.

Neutron stars are the last stable stars which can exist before the star collapses to a black hole. The formation of black holes is an inevitable consequence of the fact that gravity is an attractive and long-range force which acts upon all forms of matter and the radiation. A number of elegant theorems have been developed which demonstrate the inevitability of the formation of black holes. An intuitive picture of why black holes form can be obtained by combining the attractive nature of gravity and identity of energy and mass as expressed in Einstein's famous equation $E = mc^2$, where c is the velocity of light. Einstein's relation tells us that we can associate with each piece of mass m an energy mc^2 and, likewise, that with every piece of energy we can associate a mass E/c^2. Thus, as the star collapses, it increases its gravitational energy, which, since $E = mc^2$, is the same as adding more mass to the body. Therefore the gravitational force on the star will be stronger than before, which increases its gravitational energy and consequently its mass, and so on. In this runaway situation, the gravitational forces become stronger and stronger and are ultimately stronger than any of the other forces of physics. It is this feature of gravity which causes the formation of black holes.

To study the properties of black holes, we need a proper theory of gravity which includes Einstein's special theory of relativity, the theory which predicted exactly the relation $E = mc^2$ long before it could be tested directly by experiment. The theory of gravity developed by Einstein in the period 1910 to 1915 is called the general theory of relativity and is the best theory of gravity we possess. It is a theory of great mathematical beauty and has survived all the tests which have been made of it. In the opinion of many scientists, the construction of this theory was Einstein's greatest scientific achievement. This theory, which is of considerable mathematical complexity, is the basis for the study of black holes since it provides us with the rules for the dynamics of matter and radiation in the very extreme conditions encountered in the ultimate collapse of a star to form a black hole.

The properties of black holes are very simple. Anything which falls too close to the hole falls into it and can never come out again because gravity is so strong. This critical radius, which is technically known as the Schwarzschild radius of the black holes, is 3 kilometers for a black hole with the mass of the Sun. For black holes of other masses the critical radius is simply proportional to the mass of the hole; a black hole one million times the mass of the Sun therefore has a critical radius of 3 million kilometers. Collapse within this radius occurs for all forms of matter and radiation including light. Since light cannot escape from within the black hole to the outside, there is no direct way of observing the black hole itself. The black hole can also rotate and can have an electric charge, but these three properties—mass, rotation, and charge—are the only ones it can possess.

How, then, can one observe the phenomenon of black holes? Alice explains the procedure. It is very unlikely that matter will fall directly close to the critical radius of the black hole. Rather, the matter will possess some rotation (or more precisely angular momentum) and consequently rotational energy and it will have to lose this energy before it can come close to the black hole, fall in, and disappear forever. It is this process of losing rotational energy close to black holes which proves to be a very powerful energy source for all sorts of active nuclei. It can be shown theoretically that, in this process of falling into a black hole, a very large fraction of the rest mass energy, that is, about mc^2, of the infalling matter can be liberated. Since these events take place on a very short time scale, it can be seen that black holes have exactly the properties needed to explain the source of energy in active galactic nuclei. It can be shown that black holes are the most effective source of energy we know of for powering quasars and active nuclei, and it is highly probable that even the most extreme galactic nuclei can be explained in these terms.

There are still many problems to be understood in the theory of this process of energy generation around black holes. Obviously, the black hole only remains a powerful source of energy as long as it is supplied with infalling matter. Thus, a key question concerns the means by which gas can accumulate in the central regions of active galaxies and become fuel for the black hole. The conventional picture ascribes the source of energy to an accretion disk about the black hole in which the infalling

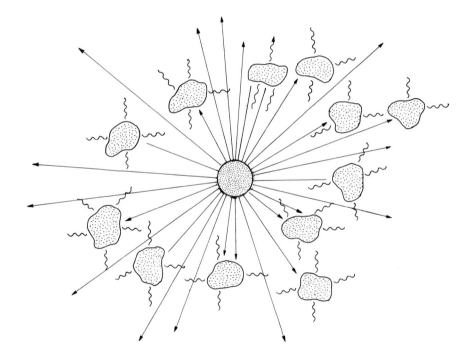

A diagram showing the likely ingredients of any plausible model
of an active galactic nucleus. There must be an active central
source surrounded by gas clouds of a range of different
densities. These clouds must have large velocities relative to the
central object. It is likely that in the most powerful objects the
disk is rather thick and that there are "funnels" out of the poles
of the black hole.

matter gradually spirals into the center, releasing its rotational energy by friction on the way in (see pp. 60 and 154). One of the most interesting questions is the origin of the fuel in different classes of active nuclei. In some cases, the gas may originate within the galaxy itself and be the result of the normal process of mass loss in stars. In other cases, the gas may result from interactions between galaxies.

The most likely candidates for the detection of black holes in astronomical systems are some of the binary X-ray sources and the nuclei of active galaxies. In the case of the X-ray binaries, which are the strongest of the X-ray sources within our Galaxy, the X-ray emission results from matter falling from the primary star onto a compact, invisible companion. Because the system is a double star, it is possible to estimate the mass of the dark companion. In a number of cases, these observations suggest that the dark companion must be more massive than the upper limit for neutron stars, about two and a half times the mass of the Sun. Because the star must be very compact, the only possible state for the dark companion is as a black hole. The X-ray emission probably originates from the inner edges of the accretion disk closest to the black hole.

The case of active galactic nuclei is described by Alice in the text for what is probably the best studied case, that of the Seyfert galaxy NGC 4151. Similar studies can be made for the other classes of active nuclei, the key features being the variability of the optical, X-ray, or ultraviolet radiation, the very large luminosities involved, and the great breadths of the strong emission lines.

5. The Caterpillar exposes a number of interesting questions concerning how theoretical astronomers tackle problems in astronomy. Should one approach an astronomical problem using accepted ideas or should one adopt a much more radical position? It is certainly safer to remain within the framework of conventional astrophysical ideas and yet the really big advances may well be made by considering much less conventional theories. What appears esoteric to one generation of astronomers often becomes the received wisdom of the next. This is one of those areas in which individual research astronomers have to judge when they have gone beyond what is likely to be a credible physical theory.

6. Technically, NGC 4151 is the nearest example of what is known as a Type I Seyfert galaxy. This means that it is one of the more extreme forms of Seyfert galaxy, in which the nucleus of the galaxy is very bright and the spectral lines are very broad. It lies at a distance of about 50 million light-years from our Galaxy.

7. The *International Ultraviolet Explorer (IUE)* is a joint United States–United Kingdom–European Space Agency satellite and has a payload consisting of an ultraviolet telescope for studying the region of the ultraviolet spectrum which cannot be observed from the surface of the Earth. The telescope mirror is 45 centimeters in diameter and the principal scientific instruments are spectrographs for studying the ultraviolet spectra of astronomical objects. *IUE* is now in its eleventh year of continuous operation in space and has been one of the most successful satellite projects ever. In many ways it can be considered a forerunner of the Hubble Space

An artist's impression of the *International Ultraviolet Explorer*
(IUE) in orbit.

Telescope, but it is much smaller in size and has a much more limited range of scientific instruments. It has had a major impact on all aspects of astronomy. Several examples of these advances have been absorbed in the text; these include the discovery of stellar winds from many classes of star, the study of accretion disks in binary systems, of active galactic nuclei, and quasars.

8. The laws of planetary motion were discoverd by Johannes Kepler in the period 1607 to 1619; these laws relate the elliptical orbits of the planets to their distances from the Sun and their velocities. These laws were interpreted by Isaac Newton, who showed that the three laws of planetary motion could be explained if it is supposed that the force of gravity is an attractive force which obeys an inverse square law and in which the strength of the force is proportional to the masses of the interacting bodies. Technically, this can be written $F = GM_1 M_2/r^2$, where M_1 and M_2 are the masses of the bodies r is their separation, and G is the gravitational constant. This is known as Newton's law of gravitation. In combination with Newton's laws of motion, these laws enable us to work out the mass of the Sun. In the case of the clouds close to the nucleus of an active galaxy, the velocities of the clouds play the same role as the rotational velocity of a planet about the Sun and an estimate of their distances from the nucleus can be made from the time delay argument.

9. For the various classes of active nuclei, see note [3].

10. HIGH-ENERGY PARTICLES AND ACTIVE GALACTIC NUCLEI

One of the most remarkable discoveries of radio astronomy has been that some active galaxies are sources of huge quantities of very high energy particles. The evidence for this comes from the radio maps of strong radio sources in which gigantic jets of radio-emitting material are observed. The most convincing interpretation of this phenomenon is that the jets consist of very high energy particles which are radiating by the synchrotron process in a magnetic field in the jet (see Chap. 5, note [3]). Some of the biggest problems of high-energy astrophysics concern the origin of these particles, why they should be ejected in jets from active nuclei, and how the beams of particles interact with the surrounding medium.

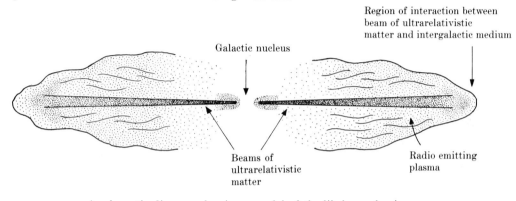

A schematic diagram showing a model of the likely mechanism for producing powerful double radio sources. A beam of high-energy particles and magnetic fields burns its way through the intergalactic gas inflating regions of high-energy particles in its wake. The radio emission observed is the synchrotron radiation of high-energy electrons gyrating in the magnetic field within these regions.

It is certain that the energy source resides in the active galactic nucleus itself and that the particles can be accelerated to very high energies there. However, exactly how this is achieved is not understood. The fact that infalling matter will form an accretion disk about a black hole provides a very nice explanation for why there should be double-sided jets associated with the release of high energy particles from active nuclei—the particles can escape most easily along the poles of the accretion disk. The interaction of these jets of particles with the surrounding medium is probably responsible for the complexities of the radio structures seen in many radio sources.

The optical emission from the nucleus itself is also likely to be due to the radiation of high energy electrons but whether or not these are related to those which form the jets seen by radio astronomers is not clear.

11. The jet in M87 was first discovered by Heber D. Curtis in 1917 but its nature was not understood at that time. It is probably closely related to the type of phenomenon seen

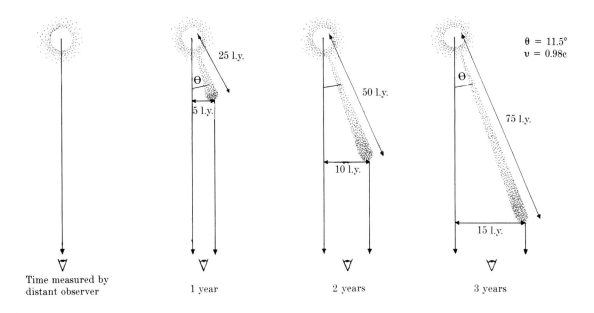

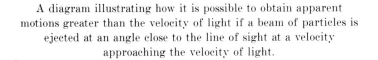

A diagram illustrating how it is possible to obtain apparent
motions greater than the velocity of light if a beam of particles is
ejected at an angle close to the line of sight at a velocity
approaching the velocity of light.

in radio sources, but in this particular case the particles in the jet are energetic
enough to radiate at optical as well as at radio wavelengths.

12. SUPERLUMINAL RADIO SOURCES

One of the most remarkable discoveries in radio astronomy in recent years has been
the discovery of very fast moving jets in the nuclei of some of the most active galaxies
and quasars. Using the technique known as Very Long Baseline Interferometry
(VLBI), radio astronomers are able to map the structures of radio sources on the
scale of $\frac{1}{1000}$ arc second or less. This enables them to probe deep into the structure of
active nuclei, although still on scales very much larger than those of the radii of
massive black holes. In the case of the quasar 3C 273, it has been discovered that the
radio lobes on this fine scale appear to be moving out from the nucleus at a speed
about 10 times the velocity of light. This phenomenon is observed in a number of
other compact radio sources—these very high velocities, apparently greater than the
speed of light, are known as *superluminal velocities*.

According to the theory of relativity, no physical object can move at a velocity
greater than the speed of light and so this phenomenon must be some sort of "optical
illusion" whereby the object appears to move faster than the velocity of light
without violating the laws of physics. It turns out that there are many ways in which

[158]

this might be done. For example, the spot on the sky illuminated by a searchlight beam can appear to move arbitrarily fast if the beam is swung rapidly enough over the sky and if the cloud is far enough away. The most likely explanation of super-luminal velocities is that the beam of radiating particles is traveling towards the observer at a small angle to the line of sight. It can easily be shown that the apparent velocity of the beam sideways can be arbitrarily fast if the beam is close enough to the line of sight and the velocity of the beam is sufficiently close to the velocity of light. This phenomenon is direct evidence for the ejection of beams of high energy particles at very high velocities from active nuclei.

8. *Big Bangs and Little Bangs*

1. COSMOLOGY

Cosmology means the study of the Universe as a whole. The word can be used to describe all aspects of the Universe including philosophical questions concerning the human situation and the ways in which different cultures have accounted for the phenomena of human existence. The word is used throughout this text in a much more limited sense in that it will refer to modern scientific cosmology, in which we attempt to understand the physical properties of everything in the Universe and the Universe itself in terms of the best physical theories we have available.

Cosmology is the subject of heated debate and controversy for a number of reasons. First, we only have one Universe to study. All scientists like to be able to repeat their experiments and try out their theories on different examples of the same type of object; in the case of scientific cosmology this possibility is totally excluded.

In addition, the laws of physics which have been thoroughly tried and tested refer to laboratory physics or to physics tested within the Solar System. It might be that there are new laws of physics which become important on a very large scale in the Universe, and we would have no other means of finding out about them except by making astronomical observations of the Universe itself. Again, for the cosmologist, this is an unsatisfactory situation.

The point of view adopted by Alice is the one adopted by virtually all cosmologists, namely, to assume that the laws of physics which have been established in the laboratory and in the Solar System are the best description of the laws which are likely to be needed to explain the large scale properties of the Universe. It turns out that a great deal can be explained using these laws, and most astronomers would argue that there is as yet no convincing piece of evidence which requires the introduction of new physics to explain cosmological phenomena. This does not mean that the laws are accepted without question. Rather it means that any piece of evidence which suggests that new physics is needed must be subjected to the most careful scrutiny. It will be shown that there are probably some questions which can only be answered on a cosmological scale (see note [17]), but this is because one is testing some of the laws far outside the regimes within which they are known to be valid.

This is the rationale behind Alice's conventional approach to the problems of cosmology.

2. THE HOT BIG BANG MODEL OF THE UNIVERSE

The Hot Big Bang Model of the Universe has become the standard framework for cosmological research for the reasons Alice gives. The Hot Big Bang model has an intriguing history. Until the present century, there was no reason to suppose that the Universe was other than stationary. The first attempt to find a self-consistent picture for the Universe as a whole was due to Einstein almost as soon as he had completed his formulation of the general theory of relativity. According to standard general relativity (and Newton's theory of gravity as well), the Universe is unstable. If we slightly squash a static Universe, the force of gravity causes it to collapse under its own self-gravitation. Einstein had no reason to suppose that the Universe was other than static and so introduced a term into the equations of motion of general relativity to produce stationary model universes.

Then, in 1929, Edwin P. Hubble discovered from observations of distant galaxies that the Universe is not stationary but is expanding. Einstein realized that he had made a grave mistake in modifying his equations to yield static model universes although there was no way he could have known otherwise at the time. In fact, the standard equations of general relativity (and Newton's theory of gravity as well) result in nonstationary models of the Universe. The exact solutions for these model universes were in fact discovered in 1922 by the Soviet theoretical physicist Aleksander A. Friedmann before the discovery of the expansion of the Universe. These models all expand from a point, and because the matter of the Universe cools as it expands, the initial state must have been very hot. This is the origin of the idea of the Hot Big Bang.

After the Second World War, George Gamow and his colleagues attempted to account for the origin of the chemical elements by nuclear reactions in the hot early phases of the Universe, but their program was only partially successful. They found that it was possible to synthesize hydrogen and helium but none of the heavier elements, such as carbon, nitrogen, and oxygen. Gamow's colleagues Ralph Alpher and Robert Hermann did however predict that a cool remnant of the hot early stages might be detectable at the present day.

This idea was forgotten until Jim Peebles and Robert Dicke began to look for this remnant in 1964. By accident, the radiation they sought was discovered in 1965 by Arno Penzias and Robert Wilson, who were commissioning a new radio antenna at the Bell Laboratories. They had discovered the *microwave background radiation,* the thermal background radiation which is the cool remnant of the hot early stages of the Universe.

At about the same time interest was revived in the synthesis of the light elements in the Hot Big Bang. It was confirmed that only the lightest elements could be synthesized in the early stages of the Hot Big Bang, but this in itself was of very great interest because it is difficult to understand how the light elements could be created in stars. The heavy elements such as carbon, nitrogen, oxygen, etc. are synthesized in the centers of stars but light elements such as lithium and deuterium (heavy hydrogen) are too fragile to survive in these very high temperature regions. The remarkable agreement between the predictions of the Hot Big Bang and the observed abundances of the light elements may be taken to be another argument in support of the Hot Big Bang model of the Universe.

3. Alice describes the three independent reasons for favoring the Hot Big Bang. It is interesting to examine the evidence for these assertions.

a. The farther away a galaxy is from us, the faster it is receding from us. A recent version of the velocity-distance relation for the brightest galaxies in clusters is shown on page 72. Because these are the most luminous galaxies known, they can be observed at great distances. The vertical axis shows the recession velocity of the galaxy away from the Earth and the horizontal axis is a measure of the distance of the galaxy from the Earth. It can be seen that there is a very strong correlation indeed, and the straight line shows what would be expected if the velocity of recession of the galaxy was exactly proportional to distance. The observations of these galaxies are very good evidence for Hubble's velocity-distance relation.

A second key feature is the fact that the Universe looks essentially the same in all directions. This is apparent from the large-scale distribution of galaxies on the sky (see p. 82), but it is even more impressive if we measure the intensity of the microwave background radiation in different directions. This is found to be the same in all directions to better than one part in one thousand, quite remarkable precision for a cosmological experiment.

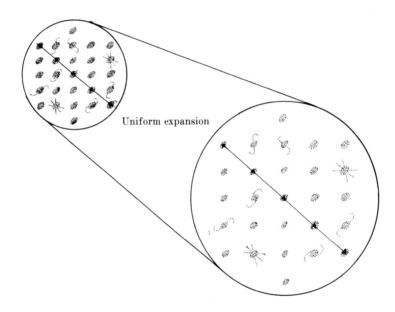

Uniform expansion

A diagram illustrating how a linear velocity-distance relation is expected in any uniformly expanding universe. In a given time interval, the whole distribution of galaxies expands. The galaxies expand away from the galaxy in which the observer is located, and the velocity of recession of each galaxy is just proportional to its distance.

These two facts tell us that the galaxies must be expanding away from one another uniformly at the present time—this is what astronomers mean by the "expanding Universe." The figure on page 162 shows a system of galaxies which expands uniformly in a given interval of time. It can be seen that if we fix our attention upon

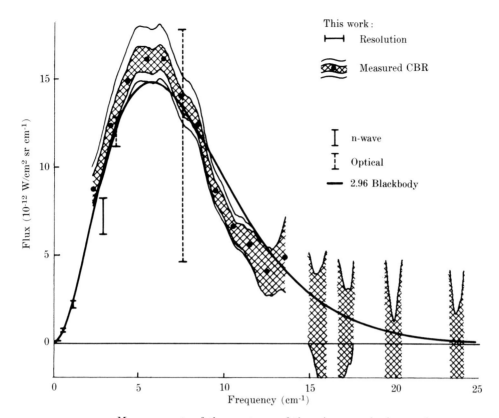

Measurements of the spectrum of the microwave background radiation. The portion of the spectrum indicated by the hatched area was obtained from a high-flying balloon. The measurements at longer wavelengths (closer to the origin of the graph) were obtained from ground-based observations. Only a few of the many measurements available at long wavelengths are shown.

any particular galaxy, during the expansion the rest of the galaxies appear to move away from it, and the further they are away, the greater their velocities of recession have to be to maintain the uniformity of the expansion.

b. The spectrum of the microwave background radiation is shown on page 163. The shaded area shows the measured spectrum and its uncertainty and the solid line shows the predicted spectrum of what is known as black body radiation at a temperature of about 2.9 K. It can be seen that the observed spectrum of the background is a very good match to the black body spectrum, particularly since there are many points which anchor the spectrum at long wavelengths which are not shown on the diagram.

This is as perfect a black body spectrum as one finds anywhere in nature. The black body spectrum is of special significance because it is the unique spectrum of radiation in thermal equilibrium. This means that all the radiation and the matter are at the same temperature. The implication of the spectral form of the microwave background radiation is that at some time the matter and the radiation in the Universe must have been in thermal equilibrium at some time in the past. Since radiation and matter cool as the Universe expands, this equilibrium occurs naturally in the early stages of the expansion of the Hot Big Bang.

c. When theorists construct the best models they can for the evolution of the early Universe, they find that, as it cools down from very high temperatures, the various elementary particles begin to combine at temperatures of about one billion degrees Kelvin to form the light elements. It turns out that the abundances of the elements predicted by these calculations depend only upon the present matter density of the Universe. These results are shown on page 75. The present density of the Universe is plotted along the horizontal axis and the abundance of the light elements along the vertical axis. Also shown on the graph are some estimates of the present observed abundances of these light elements. It is possible to account rather satisfactorily for the observed abundances of helium, deuterium, and lithium in low-density models of the Universe. This is a remarkable result. This process of creating the light elements is successful because the synthesis of the elements occurs in a rapidly cooling expansion and is completed within the first 15 minutes of the life of the Universe. There is not time for the heavier elements like carbon, nitrogen, and oxygen to form from the light elements. This process contrasts with heavy element synthesis in stars which takes place over time scales of millions or billions of years.

4. Tweedledum and Tweedledee meet Alice in chapter 4 of *Through the Looking-glass.* According to the nursery rhyme,

> Tweedledum and Tweedledee
> Agreed to have a battle;
> For Tweedledum said Tweedledee
> Had spoiled his nice new rattle.

It seems appropriate that Tweedledum and Tweedledee should take opposite sides on all the cosmological problems Alice lists.

5. Hubble's constant H_0 is the parameter which describes how rapidly the Universe is expanding at the present time. It is the constant in the velocity-distance relation,

$$\text{Recessional velocity} = \text{Hubble's constant} \times \text{distance}$$

or $$v = H_0 r$$

The velocity is measured in kilometers per second ($km\ s^{-1}$) and the distance in megaparsecs (Mpc), which is a convenient astronomical distance measure. One megaparsec is roughly three million light-years (see note [1], Chap. A6).

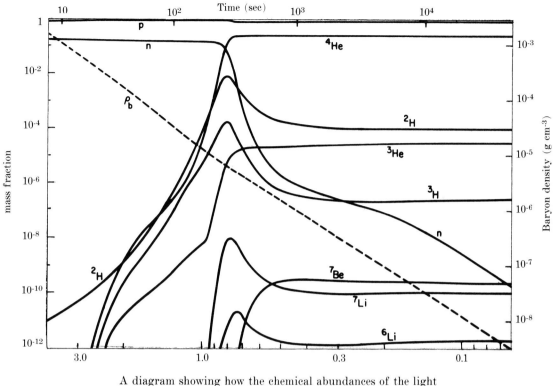

A diagram showing how the chemical abundances of the light
elements are expected to change in the first 30 minutes of the
Hot Big Bang. This is only one of many runs made by Dr.
Robert Wagoner of computer program which work out the
synthesis of the elements in the Hot Big Bang.

Hubble's constant also determines the age of the Universe. If the Universe had expanded at its present rate from the very beginning to now, its age would simply be the inverse of Hubble's constant; that is, for an undecelerated Universe,

Age of Universe = 1/Hubble's constant = $1/H_0$.

Using the values given by Tweedledum and Tweedledee, if $H_0 = 100$ km s^{-1} Mpc^{-1}, the age of the undecelerated Universe is 10 billion years. If $H_0 = 50$ km s^{-1} Mpc^{-1}, the age of the undecelerated Universe is 20 billion years. In a real Universe, there must be some gravitational deceleration of the expansion and so the Universe must have expanded more rapidly in the past. Therefore all real world models should have ages which are less than the values quoted above. The greater the deceleration, the greater will be the difference.

This uncertainty of a factor of two in the measured value of Hubble's constant is attributable to the difficulty of measuring accurate distances for distant galaxies. One needs good distance indicators which are independent of the redshift of the galaxy.

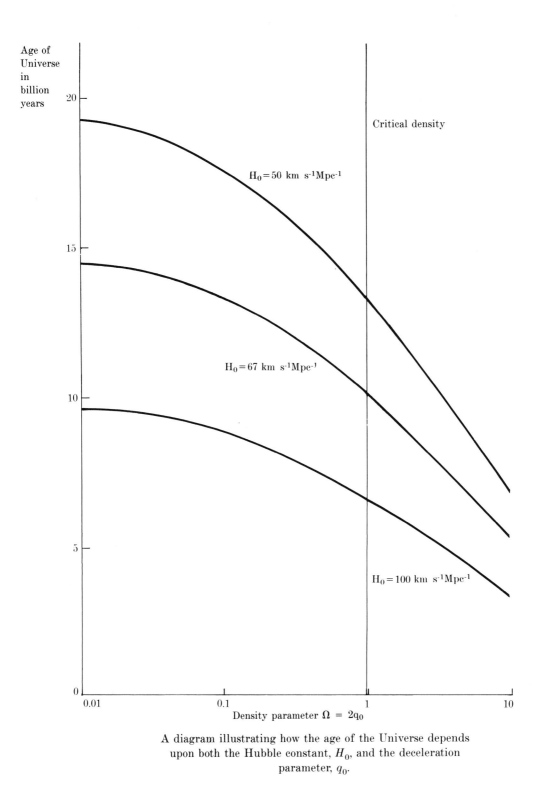

A diagram illustrating how the age of the Universe depends upon both the Hubble constant, H_0, and the deceleration parameter, q_0.

6. The measurements of the velocity of light and their errors are taken from *The American Handbook of Physics*, 2d ed. (New York: McGraw-Hill, 1969).

7. Cepheid variable stars are among the best distance indicators for nearby galaxies. There is a very strong relation between the luminosity (i.e., total energy output) of these variable stars and the period of regular variation in their brightnesses. Therefore, if one can measure the period of the variations in brightness of one of these stars in a distant galaxy, its luminosity can be inferred from the period-luminosity relation. Then, by measuring the observed brightness of the star, its distance can be derived. The Cepheid variable stars have two big advantages as distance indicators. First, they are very luminous stars and so can be observed far away. Second, they have a characteristic variation of luminosity with time and so can be detected with confidence, even if they are very faint. As Alice notes, this will certainly be one of the most important programs for the Hubble Space Telescope.

8. CLASSICAL COSMOLOGY

Alice describes one of the great classical problems of cosmology. Aleksander A. Friedmann was the first to discover the solutions of Einstein's equations of general relativity for uniformly expanding universes. In fact, most of the essential features of these models can be derived from a simple Newtonian model for the expanding Universe. It turns out that we can model the present expansion of the Universe by the expansion of a uniform sphere of matter in which the only force present is the attractive force of gravity. The sphere expands uniformly because we know we have to construct uniformly expanding model universes. The only force acting upon the

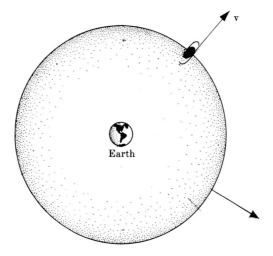

A simple model for working out the dynamical behavior of an expanding Universe.

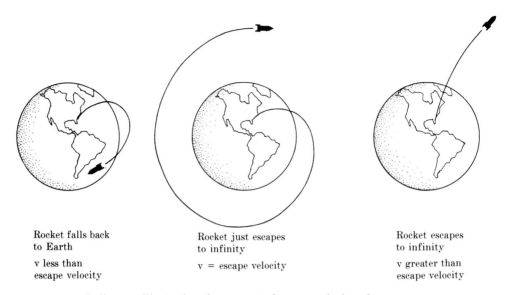

Rocket falls back
to Earth

v less than
escape velocity

Rocket just escapes
to infinity

v = escape velocity

Rocket escapes
to infinity

v greater than
escape velocity

A diagram illustrating the concept of escape velocity of a space
vehicle from the surface of the Earth.

expanding sphere is the attractive force of gravity which slows down its expansion.
We can understand the behavior of the sphere if we consider the force acting upon a
galaxy which is located at the edge of the expanding sphere and which expands with
the sphere. Clearly it will be slowed down because of the gravitational attraction of
the rest of the sphere.

This situation is analogous to the problem of the escape velocity of a rocket from
the surface of the Earth. If the rocket is traveling fast enough, it can escape from the
pull of the Earth's gravitational field. If it is not traveling fast enough, it will not
escape from the Earth and will fall back to the Earth's surface. There is a certain
critical velocity at which the rocket can just escape from the Earth and no more—it
ends up at infinity with zero velocity. This critical velocity is known as the *escape
velocity.*

The dynamics of uniformly expanding models of the Universe are exactly the
same (see pp. 74 and 169). If the Universe expands fast enough to begin with, it can
expand to infinity and end up there with a finite velocity. If it does not expand fast
enough, it will eventually stop expanding and collapse back to a hot dense state
again. There is a critical model which separates these two types of behavior in which
the Universe just expands to infinity and stops there. Which type of behavior is
actually found depends upon how great the deceleration of the Universe is. According to Newton's law of gravitation and Einstein's general theory of relativity, the
deceleration is due entirely to the matter present in the Universe. The greater the
density, the greater the deceleration. There is therefore a critical density for the
Universe which separates those models in which the Universe will expand forever
from those in which the Universe will eventually collapse back upon itself. This
density is known as the *critical density* and naturally its value depends upon the
present expansion rate of the Universe.

[168]

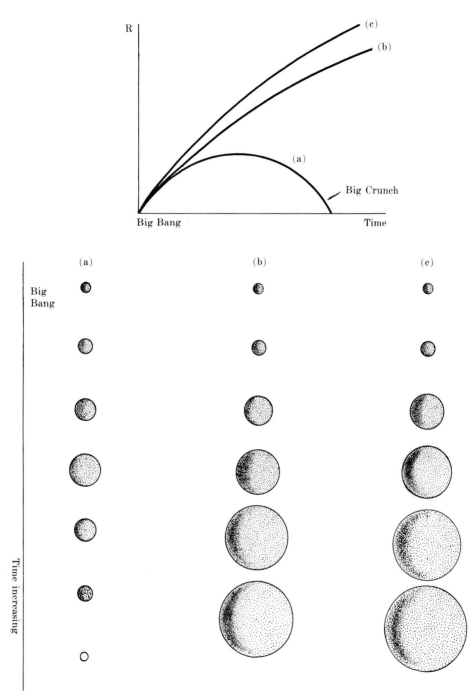

A comparison of the dynamics of different models of the
expanding Universe: *(a)* a high-density model, which eventually
collapses to a 'Big Crunch'; *(b)* the critical model, in which the
Universe can just expand to infinity; and *(c)* a low-density
Universe, in which the Universe is open and expands to infinity.

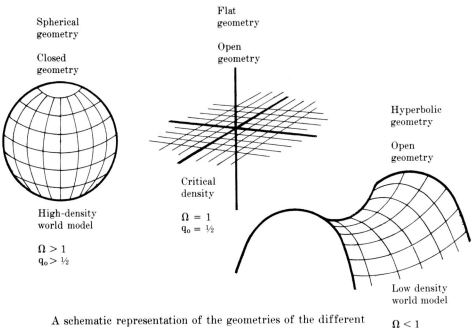

Spherical
geometry

Closed
geometry

High-density
world model

$\Omega > 1$
$q_o > \frac{1}{2}$

Flat
geometry

Open
geometry

Critical
density

$\Omega = 1$
$q_o = \frac{1}{2}$

Hyperbolic
geometry

Open
geometry

Low density
world model

$\Omega < 1$
$q_o < \frac{1}{2}$

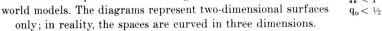

A schematic representation of the geometries of the different
world models. The diagrams represent two-dimensional surfaces
only; in reality, the spaces are curved in three dimensions.

We can therefore characterize the dynamics of the Universe either by its present
deceleration or by its density. In both cases there are critical values for these
parameters, and these are traditionally chosen so that the actual density of the
Universe ρ_o is compared with the critical density ρ_{crit}. The density parameter Ω is
defined to be the ratio of these densities $\Omega = \rho_o/\rho_{crit}$ so that if Ω is greater than 1, the
Universe collapses and if Ω is less than 1, it expands forever. The deceleration
parameter q_0 is a dimensionless number which describes how rapidly the Universe is
decelerating. Its critical value is $\frac{1}{2}$. If the value of q_0 is greater than $\frac{1}{2}$, the Universe
will eventually collapse, and if q_0 is less than $\frac{1}{2}$, the Universe will expand to infinity.
For the models of the Universe of general relativity, $\Omega = 2\,q_0$. This relation provides
an important test of the validity of these models.

One other miraculous piece of physics associated with these models of the Universe
comes directly out of the general theory of relativity: the geometry of space itself is
most unlikely to be flat. In our three-dimensional world we are used to the fact that
the angles of a triangle add up to 180 degrees and that parallel lines only meet at
infinity. It was one of Einstein's great insights to realize that in the presence of
gravitational fields the geometry of space is not flat but curved. In fact, not only is
space curved but space-time itself, the four-dimensional generalization of three-
dimensional space which includes time as well as spatial dimensions, is also curved.
We need not go into the complications of general relativity but we can make one
simple point: the three types of dynamical behavior shown on pages 74 and 169 are
closely related to the different possible geometries of the Universe (see p. 170).

If the Universe expands forever, that is, if q_0 is less than ½ and Ω is less than 1, the geometry of the Universe is hyperbolic. If the Universe ultimately collapses, that is, if q_0 is greater than ½ and Ω is greater than 1, the geometry of the Universe is spherical. Only in the case of the critical model, $q_0 = $ ½ and $\Omega = 1$, is the geometry of the Universe flat in the sense that the spatial geometry obeys the rules of Euclidean geometry. All others have non-Euclidean geometry.

The pioneers of observational cosmology hoped that, by making observations of very distant galaxies, it would be possible to distinguish which type of Universe we live in. We can understand simply why it is that the observed properties of distant galaxies must depend upon the deceleraton of the Universe. How far away a galaxy was when it emitted the light we receive now must depend upon how rapidly the Universe has been decelerating. Therefore, the observed properties of distant objects depend upon the model of the Universe and, if we can measure the physical properties of objects far away and compare them with their observed properties, we should be able to find out the deceleration of the Universe. I refer to the attempts to determine the basic parameters of the Universe H_0, q_0, Ω, and its present age as *classical cosmology.*

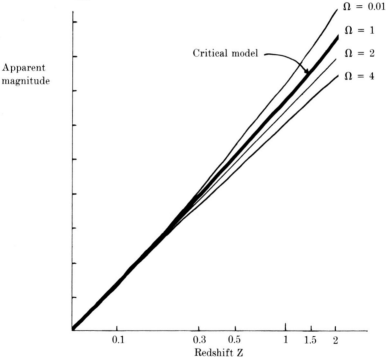

Illustration of how the redshift-apparent magnitude relation for a standard object is expected to vary for different world models. By measuring the shape of the relation it is possible, in principle, to measure the deceleration of the Universe. This diagram shows the expected relations for sources with simple power-law spectra in which luminosity is proportional to the inverse of frequency ($I \propto \nu^{-1}$).

This turns out to be a very difficult astronomical problem. We do not understand the properties of galaxies and other objects well enough to be able to derive accurate estimates of the present deceleration of the Universe from observations of these distant objects. Realistically speaking, the deceleration parameter q_0 probably lies somewhere in the range from 0 to 1, but many astronomers would even consider this statement to be unduly optimistic.

In this situation, many astronomers appeal to arguments of a more astrophysical nature to support their preferred values for the deceleration parameter. Tweedledum and Tweedledee give a reasonable summary of the types of argument which are employed. The main points are as follows:

a. The Universe must be older than the oldest objects we know of in the Universe, namely, the globular clusters. Their ages may be estimated from the theory of stellar evolution. This involves fitting the observed temperature-luminosity diagrams to theoretical models of the evolution of stars from the main sequence onto the giant branch. The ages derived in this way lie in the range 13 to 17 billion years.

Tweedledum says that this supports his position because his value of $1/H_0$ is about 20 billion years and there cannot be much deceleration or else the age of the Universe will become less than the ages of the globular clusters (see p. 166). Tweedledee says that he is not so confident that the ages of globular clusters have been measured so accurately. After all, this age does depend upon understanding the physics of stellar interiors and we still have the solar neutrino problem to worry about (see note [1], Chap. A4).

b. The models for the synthesis of the light elements in the early Universe suggest that the density of ordinary matter in the Universe must be about one-tenth the critical density. Tweedledum argues that this means that the Universe must expand forever and that q_0 is less than ½. Tweedledee maintains that this argument results in only a lower limit to the total amount of matter present in the Universe. We can never discover *less* matter than we know of now. As explained by the Dormouse (see chap. 5, note [8]), there could be lots of matter present in less normal forms, and we have yet to solve the problem of the hidden mass. For example, it is theoretically possible to build model universes in which there is a low average density for the ordinary matter and the critical density for some form of very weakly interacting unknown particles. In this case the Universe would have $q_0 = $ ½ and the products of primordial nucleosynthesis would be the same as in a low-density universe.

c. One interesting argument which is not strictly scientific but which should not be neglected is that the Universe must be within a factor of about 10 of the critical values for the density and deceleration parameters. If we consider setting up Universes at random, there is no a priori reason to pick any particular values of H_0, q_0, or Ω which would end up with our present Universe 10 or 20 billion years after the beginning. Some scientists argue that, since q_0 and Ω could take any values between zero and infinity, the only reasonable non-zero choice for the Universe would be the critical value. Some support for this value comes from recent developments in the theory of elementary particles as applied to the very early Universe. We will have to *wait* and see how well these ideas survive further scrutiny.

One approach to this problem is through what is known as the *anthropic principle*. According to this principle, certain preconditions must be fulfilled before observers can exist who are capable of observing the Universe. For example, if life such as our own is to develop, there must have been time for biological molecules to form. Equally, the constants of physics have to be such that things like stars, planets, and galaxies have time to form. The proponents of these ideas point out that one can make the case that only in universes similar to our own would there exist the possibility of life forming. This may "explain" why Hubble's constant corresponds to universes with ages about 10 to 20 billion years and values of q_0 and Ω within about a factor of 10 of the critical values. In other words, the needs of biological evolution select the only Universe in which observers can exist.

Most astronomers would regard the anthropic principle as a last resort if all other explanations of the present state of the Universe fail. Most of us would prefer a strictly physical explanation for the present properties of the Universe and values of the fundamental constants of nature.

9. THE AVERAGE DENSITY OF THE UNIVERSE

The above exposition of classical cosmology makes a very clear presumption that the deceleration of the Universe is due entirely to the matter present in the Universe. Formally, this result has been written in terms of the relationship between the deceleration and density parameters $q_0 = \frac{1}{2}\Omega$. This relation can be tested by astronomical observations because we can measure separately the deceleration of the Universe and its average density and so test the underlying theory by observations on the scale of the Universe itself.

We described above the procedures for finding the deceleration parameter through observations of distant galaxies and other objects. We can also estimate the density of the Universe by measuring the amount of mass it contains on as large a scale as possible. We described in note [7] of Chapter 6 how masses are estimated in astronomy. These same techniques can be used on large scales in the Universe where it is likely that one is taking averages over all forms of visible and invisible matter. It should be recalled that the force of gravity is universal and affects all forms of matter, radiation, and energy in the same way. These mass estimates show that the Universe must be within about a factor of 10 of the critical density. Typically the values found for the density parameter Ω lie in the range 0.1 to about 0.6, that is, formally the values are less than the critical value with a best estimate of approximately 0.3. This value is somewhat larger than the values found from the argument concerning primordial synthesis of the elements, but not by a large amount.

In many ways, the most important aspect of these estimates of Ω and q_0 is that they agree reasonably well with what is expected for the standard models of the Universe. If new forces of physics were to appear only on a large scale in the Universe, one might have expected there to be a gross discrepancy between the values of q_0 and Ω. What Alice claims is that since we can state with some confidence that $q_0 = \frac{1}{2}\Omega$ within a factor of about 10, it is likely that the equality is exact and that the standard model is correct. However, we would like to know this with much greater precision— say, 10 times greater precision—than is currently available. We would then have confidence that we have at least measured the total mass of all the constituents of the Universe.

10. The classical models of the Universe of general relativity describe the overall structure of the Universe on the very largest scale. They smooth out all the real structure in the Universe and have nothing to say about how galaxies came about or how they have evolved into their present state. This study is the subject of *astrophysical cosmology*—the observational and theoretical study of how all the objects in the Universe came into being and how they have evolved with time. This is one of the areas in which there has been great progress in recent years and one which is likely to be among the most productive for studies with the Hubble Space Telescope.

11. REDSHIFT AND TIME IN COSMOLOGY

When we receive light from objects at very great distances, we are looking back into the past to times when the Universe was much younger than it is now. To understand how astronomers measure time and distance in cosmology, we must understand the meaning of the *redshifts* of galaxies and quasars.

We are familiar with the phenomena of redshifts and blueshifts in our everyday life. If a police car rushes towards us with its siren sounding, the pitch of the note it emits is higher than it would be if the car were stationary. Likewise, if the car rushes away from us, the note we hear is of a lower pitch than if the vehicle were at rest. What is happening is that the wavelength of the radiation is squashed or stretched depending on the motion of the source of the sound relative to the observer. In the case of the car moving away from us, the wavelength is stretched; exactly the same phenomenon occurs for the light from galaxies which are receding from us in the expanding Universe (see p. 175). Since the light is shifted toward the red end of the optical spectrum, this phenomenon is referred to as the redshift of galaxies.

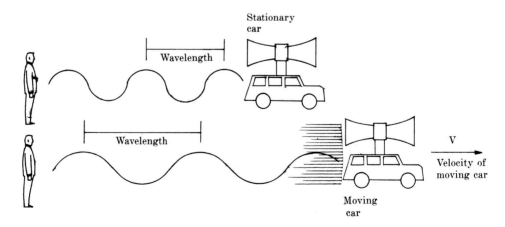

Illustration of the effect of motion on the observed frequency of the note emitted by a stationary and a moving vehicle. This phenomenon is known as the Doppler effect.

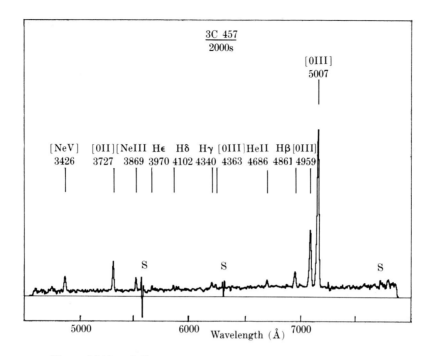

The redshifts of distant galaxies. The spectra shown are those of the radio galaxies 3C 457 (redshift 0.428) and 3C 22 (redshift 0.938). The same emission lines are observed in both spectra, for example [0II], [0III], and [NeV], but in the distant galaxy they are all shifted to the red end of the spectrum. The rest wavelengths of the lines are shown above each spectral line.

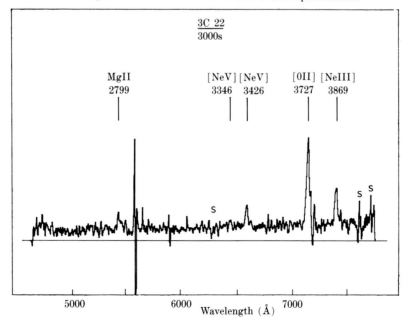

For galaxies and quasars, the redshift is defined to be the stretching in wavelength relative to the emitted wavelength, that is,

$$\text{Redshift} = z = \frac{\lambda_{Obs} - \lambda_{em}}{\lambda_{em}}$$

where λ_{em} and λ_{obs} are the emitted and observed wavelengths of the radiation, respectively. We can show that when the velocity of recession of the galaxy is very much less than the velocity of light, the redshift is simply related to the velocity of recession of the galaxy by $v = cz$ where c is the velocity of light (see p. 72).

In cosmology, however, the redshift has a more profound meaning. If we form the quantity $R = 1/(1 + z)$, this quantity R tells us how large the Universe was when the light was emitted relative to its present size. R is therefore called the *scale factor* of the Universe and is the quantity plotted on the vertical axis of the figures on pages 74 and 169; it shows the expansion of the Universe for different values of the deceleration and density parameters. For example, nearby objects have essentially zero redshift and hence $R = 1$, that is, this is the relative size of the Universe now. When we observe a galaxy at a redshift $z = 1$, the light was emitted from the galaxy when the scale factor was $R = 1/(1 + z) = \frac{1}{2}$, that is, the Universe was half its present size. If we observe a quasar with a redshift of 4, the relative size of the Universe was only one-fifth its present size. Thus, the redshift is actually a measure of the size of the Universe.

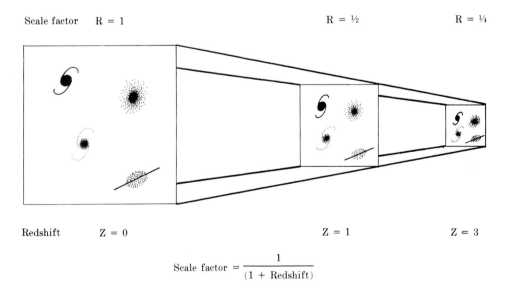

Scale factor R = 1   R = ½   R = ¼

Redshift   Z = 0   Z = 1   Z = 3

$$\text{Scale factor} = \frac{1}{(1 + \text{Redshift})}$$

Illustration of the meaning of the scale factor and of redshift in cosmology. It can be shown that in the simple uniform models of the Universe, scale factor = 1/(1 + redshift).

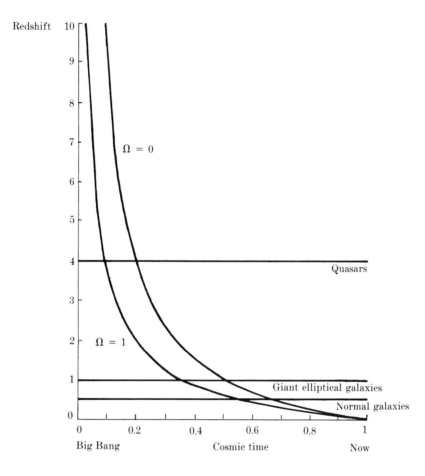

The relation between time and redshift for different cosmological
models. Also shown are the largest redshifts at which different
classes of object have been observed from the ground.

The catch is that we do not know when the light was emitted. To find this out, we
have to use the models of the Universe described in note [8], each of which makes a
definite prediction about the relation between the scale factor R and time in the
Universe. For example, let us consider an empty Universe, in which there is no
deceleration, $q_0 = 0$, and no matter present, $\Omega = 0$. In this case, the scale factor
increases exactly proportional to time; therefore, when the scale factor was ½, the
Universe was half its present age; at a redshift of 4, the Universe was one-fifth its
present age, and so on. In all the other models, a given redshift corresponds to earlier
times than those found in the undecelerated model.

The redshifts to which different classes of object have now been observed are also
shown on page 177. We can observe luminous galaxies back to roughly half the
present age of the Universe and the brightest quasars to epochs when the Universe

was one-fifth of its present age or younger. This diagram illustrates how, when objects of very large redshift are considered, it is much more useful to think in terms of time rather than distance. Note, however, that to convert redshifts into times, we have to make assumptions about the type of Universe we live in. The redshift is an unambiguous measure and is the best of all measures to use. To give it a physical meaning, simply convert it into the scale factor $R = 1/(1 + \text{redshift})$ and then R is the size of the Universe relative to its present size when the light we now observe was emitted.

12. THE COSMOLOGICAL EVOLUTION OF RADIO SOURCES AND QUASARS

One of the big surprises of the last 25 years in extragalactic astronomy and cosmology has been the discovery that there were many more active galaxies and quasars in the past than there are today. This phenomenon is referred to as the cosmological evolution of radio sources and quasars—evolution in the sense that their bulk properties change with time in the Universe.

The evidence for this was discovered in the late 1950s and the 1960s, when the first complete sky surveys of radio sources were made. The radio sky differs dramatically from the optical sky. In the optical picture of the sky, the brightest objects are stars, which are nearby objects in our own Galaxy. In contrast, at radio wavelengths, the brightest objects in the sky are very distant objects indeed. This makes a great deal of sense because the distribution of these objects in the sky is remarkably uniform. This was confirmed dramatically in 1962, when one of the brightest radio sources, 3C 273, was found to be a quasar with a large redshift (see Chap. 6, note [3]).

It was therefore natural to assume that when the radio surveys were extended to faint radio limits very distant objects would be discovered. This has indeed proved to be the case, but what was not expected was that there would be so many more active galaxies and quasars than we observe nearby. Notice that this statement refers to the relative probability of radio source and quasar activity occurring. The increase in the observed space density of these objects in the past is over and above the increase expected because of the expansion of the Universe. When the Universe was about a third or a quarter its present age, there were roughly 1000 times more quasars and radio sources than there are now (see p. 78). The origin of this phenomenon is not understood, largely because we do not understand the astrophysics of quasars and radio sources well enough. However, these data provide us with definite evidence that the Universe went through a phase of very great activity in the relatively recent past. One of the great challenges is to understand why this occurred.

13. As usual, Tweedledum and Tweedledee adopt opposing views on the origin of galaxies and clusters of galaxies in the expanding Universe. The immediate question which spurs this argument is whether the first galaxies formed when the Universe was very young or in the relatively recent past. Tweedledum speculates that there may have been a large amount of activity associated with active galaxies and quasars

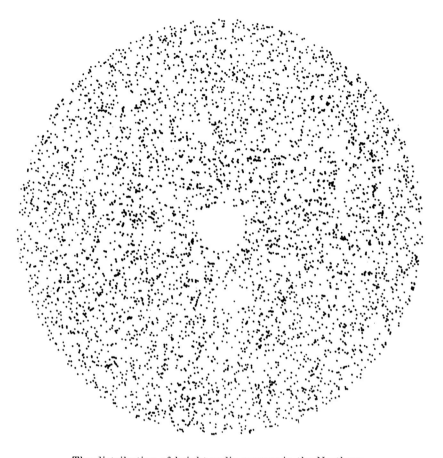

The distribution of bright radio sources in the Northern
Hemisphere. The coordinates have been squashed so that equal
areas are properly represented. The hole in the center of the
diagram occurs because the area around the North Celestial Pole
was not surveyed.

when the Universe was about a quarter its present age because galaxies and clusters
of galaxies only formed at this relatively late stage in the Universe.

One intriguing piece of evidence which has recently come to light is that although
there is a huge increase in the number of quasars at redshifts of about 2 and 3
compared with the numbers nearby, it is very difficult to find them at redshifts
greater than about 4. Only six quasars are known with redshifts greater than 4, the
most distant object being the quasar 0051-279, which has redshift 4.43. One interpretation of this phenomenon is that the host galaxies were just forming at this epoch
in the Universe. There are, however, a number of other plausible explanations for the
lack of quasars at redshifts greater than 4 and Tweedledee is not slow to point them
out.

14. THE FORMATION OF GALAXIES

The origin of galaxies and clusters of galaxies is one of the most important and yet difficult problems of modern astrophysical cosmology. Gravity has the tendency to cause regions of the Universe to collapse unless they possess some form of internal support. This is helpful in understanding why matter should be clumped into objects like stars, galaxies, and clusters, but there are problems when we repeat this calculation for matter in an expanding Universe. Evgeni M. Lifshitz first showed in 1946 that, although density inhomogeneities grow in an expanding Universe, they grow very slowly. According to our present understanding of gravitational collapse in the Hot Model of the Universe, we can only make galaxies and clusters if we include finite inhomogeneities in the matter distribution in the early Universe to begin with. This is rather unsatisfactory because it means that we must put in the correct conditions at the beginning in order to obtain galaxies at the present day. A possible solution to this problem is described by Alice at the end of this chapter, but it must be classified as speculative at the moment.

Despite this unsatisfactory situation concerning the initial conditions from which the Universe expanded, it is possible to work out how we would expect these initial inhomogeneities to evolve as the Universe expands. This has resulted in two basic types of behavior and two quite different pictures of how structure came about in the Universe. Needless to say, Tweedledum and Tweedledee adopt these opposing positions.

In one picture, known as the *adiabatic model*, the inhomogeneities originate as sound waves in the matter and radiation of the early Universe. When the matter and radiation in the Hot Model cool below about 4000 K at a redshift of about 1500, the ionized matter recombines to form a neutral gas and these inhomogeneities begin to collapse to form bound objects. In this picture, all the small-scale inhomogeneities are damped out before this time and so the first things which collapse to form real structures are the very largest systems, such as clusters of galaxies, and even larger structures, such as superclusters and the stringy structure seen on pages 82 and 83. These systems can only be formed late in the history of the Universe, when the average density of the Universe is similar to that of the large-scale systems we now observe. Tweedledum likes this model because it can explain why it is physically reasonable that galaxies and clusters should form late in the Universe at redshifts of, say, 5.

There is another successful aspect of this model. When very large inhomogeneities begin to collapse, they do so most rapidly along their shortest axis. This means that they first collapse into planes which are often called pancakes. Contraction and fragmentation continues within the pancakes and it is out of these fragments that galaxies form (see pp. 80 and 81). The clustering of galaxies within these sheets and filaments produces groups and clusters of galaxies. This overall picture can explain rather neatly the observed large-scale distribution of galaxies in the Universe, which seems to possess a filamentary or cellular structure (see pp. 82 and 83).

The other main class of models is referred to as *isothermal models*. In these, the perturbations in the early Universe are only in the matter component and not in the radiation, that is, rather than sound waves, they are just lumps in the overall

Adiabatic fluctuations

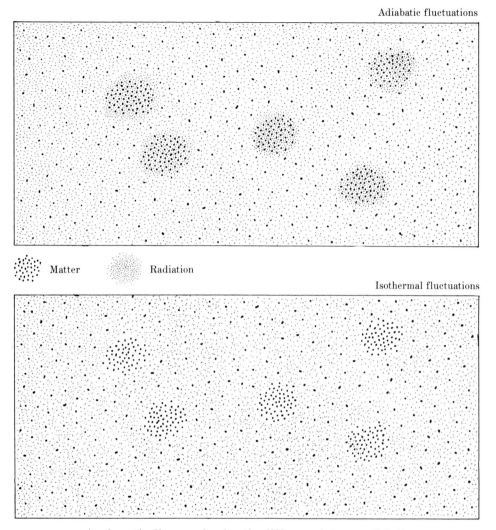

Matter   Radiation

Isothermal fluctuations

A schematic diagram showing the difference between adiabatic
and isothermal fluctuations during the early radiation-dominated
phases of the Universe.

distribution of matter. These are not damped out as the sound waves are, but all scales
of irregularity can survive to a redshift of about 1500 when the Universe becomes
transparent to radiation and collapse of the irregularities begins. This means that
low-mass objects can first form soon after this redshift. The objects most commonly
formed have masses approximately one million times the mass of the Sun, similar to
the masses of star clusters we observe in our Galaxy today. In this picture, galaxies
and clusters form by a process of hierarchical clustering so that the star-cluster-
sized objects cluster into galaxies, the galaxies into clusters of galaxies, and so on as
the Universe grows older (see pp. 80 and 81).

These two models result in quite different predictions about when galaxies could first have formed. In the adiabatic model, galaxies must have formed only late in the Universe, whereas, in the isothermal model, they could have formed much earlier. One important point mentioned by Tweedledee is that so far as one can tell, the most distant quasars seem to possess roughly the same abundances of the heavy elements as nearby objects. Since the heavy elements are created in stars, this means that there must have been considerable amounts of star formation within the parent galaxies by a redshift of 3. It is not clear whether or not this poses a problem for models in which galaxies form late in the Universe. Some authors have proposed that there may have been an early generation of "pregalactic" stars to make the heavy elements we see in the quasars and also in some of the very oldest star systems in our own Galaxy.

Of course, we can devise alternative scenarios which incorporate the best features of both models. In addition, the models have been developed to take account of the possibility of unknown weakly interacting particles and massive neutrinos. However, the basic features of these pictures remain more or less the same.

15. INTERGALACTIC GAS

The quasars act as excellent probes for any gas which lies along the line of sight between the quasar and the Earth, because any gas along the line of sight absorbs radiation at the resonant frequencies within atoms—which are also the cause of the strong resonance emission lines. Such absorption lines are frequently observed at optical wavelengths in quasars at great distances because the strong resonance lines of the common elements are redshifted into the optical region of the spectrum.

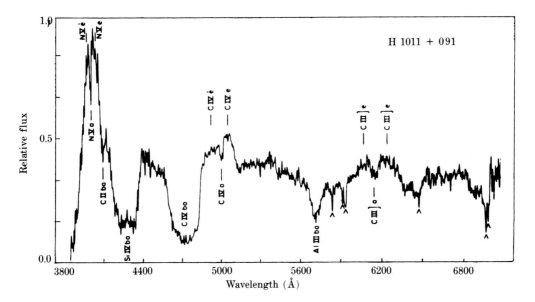

The spectrum of the quasar H 1011 + 091 showing the broad
absorption troughs associated with the strong emission lines,
suggesting that the material is ejected from the quasar.

There appear to be at least four different causes of the absorption lines we observe in quasars.

a. There are deep absorption troughs to the short wavelength side of strong resonance emission lines. These are almost certainly caused by gas flowing out from the quasar at high velocities, the blueshifting of the absorption features being caused by the movement towards the observer of the ions or atoms responsible for the absorption.

b. In some cases, there are several rather sharp absorption lines seen very close to a strong, broad emission line in the quasar's spectrum. It is highly probable that these are caused by galaxies in the vicinity of the quasar, possibly galaxies belonging to the same cluster or group of galaxies as the quasar.

c. In some cases, there are series of lines of the common elements which correspond to the same redshift and are similar to those expected in an intervening galaxy along the line of sight to the quasar but at a very different redshift. These are chance events which occur rather frequently in the spectra of very distant quasars.

d. Some systems appear to consist only of the absorption lines of hydrogen. These have only been discovered so far in very large redshift quasars and result in a "forest" of absorption lines to the short wavelength side of the strong Lyman-α emission line of hydrogen (see pp. 84 and 85). These are probably diffuse intergalactic clouds of primordial material which have not condensed into galaxies or clusters.

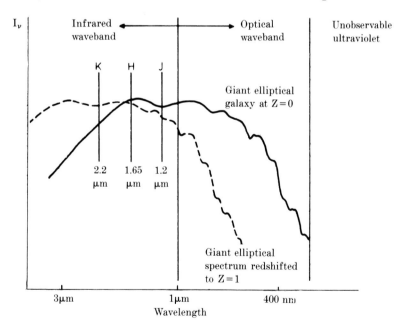

A diagram illustrating how the light of a giant elliptical galaxy is shifted into the infrared waveband when it is observed at a large redshift.

These systems are of special interest and importance because they may well contain unprocessed matter, containing the primordial abundances of the elements synthesized in the Hot Big Bang.

Tweedledum and Tweedledee argue about the likely origin of the absorption features seen in quasar spectra. To some extent, both of them must be right.

16. The reason for observing distant galaxies in the infrared waveband is twofold. First of all, at redshifts greater than 1, most of the optical emission is redshifted into the infrared waveband. Galaxies are therefore much more readily observed in the infrared waveband than in the optical waveband at very large redshifts. Second, the types of star which contribute most of the luminosity of a galaxy in the infrared part of the spectrum are red giant stars and are formed from the old stellar populations of galaxies. They are therefore much less sensitive to bursts of star formation, which can strongly influence observations made in the optical waveband.

The result of this second factor is that it is remarkably easy to predict the change in luminosity of a galaxy with time in the infrared waveband since it is attributable to the different numbers of stars evolving onto the giant branch at different cosmic epochs. Our observations suggest that we observe more or less exactly the amount of evolution expected if the underlying stellar population of the galaxy formed at a large redshift and if that population has simply aged with time, the total luminosity of the galaxy growing fainter with time.

17. THE BASIC PROBLEMS OF COSMOLOGY

Alice gives a brief survey of some of the basic problems of cosmology and how recent developments in elementary particle physics may help resolve some of them. These have been touched upon briefly but it is useful to summarize them.

a. The matter-antimatter asymmetry. We know that our Universe is made out of matter rather than a mixture of matter and antimatter and yet, theoretically, in the very early Universe, there must have been almost equal numbers of particles and antiparticles, far outnumbering the particles of matter we see today. This is because there must have been roughly a particle-antiparticle pair in the very early Universe for every pair of photons (or particles of radiation) we see in the microwave background radiation today. We know that the photons of the microwave background radiation outnumber the particles of matter in the Universe today by a factor of approximately 100 million; hence there must have been a one-part-in-100-million asymmetry in favor of matter rather than antimatter in the very early Universe.

b. We described in note [14] how we have to put in finite-sized perturbations at the beginning of the Universe to ensure that we will make galaxies and clusters by the present day. Where did these irregularities or inhomogeneities come from?

c. The Universe looks the same in all directions today, but how did it know that it had to end up so uniform on the large scale when it began? It can be shown that as we go farther and farther back in time, we can communicate less and less far in the time available since the expansion began. How was it that one bit of Universe was able to communicate with another bit so that they end up having the same average properties today?

According to classical cosmology, these three properties have to be built into the initial conditions from which the Universe expanded. This is a totally unsatisfactory situation because it means that, in order to explain the Universe today, we have to put in exactly the correct conditions at the beginning and, to any scientist, this means we have explained nothing—we only get out at the end what we put in at the beginning.

Alice points out, however, that this unsatisfactory situation may be resolved through some remarkable recent developments in the theory of elementary particles and of Grand Unified Theories to explain the forces of nature. These theories predict that, although the very early Universe was symmetric, small but finite asymmetries between matter and antimatter can develop at later stages. This would account for the asymmetry between matter and antimatter in the Universe. Likewise, there may be associated ways of producing the initial spectrum of fluctuations at these very early epochs, and even the isotropy of the Universe may be plausibly explained by these theories. A nontechnical description of some of these exciting ideas is given in the book *The Left Hand of Creation,* by Drs. John Barrow and Joseph Silk (New York: Basic Books, 1983).

Tweedledum and Tweedledee express their doubts about what astronomers can do about many of the highly technical questions raised by these new ideas. The most important thing astronomers can do is provide the theorists with as strong constraints as possible upon the initial conditions from which the Universe must have evolved. In turn, these observations can constrain a number of aspects of the theory of elementary particles. Whatever the outcome, it now seems inevitable that the concerns of the cosmologist and the elementary particle physicist will ultimately be intertwined in an effort to understand what is perhaps the most profound of all physical problems—the origin of the Universe itself.

18. I have expressed my own prejudices in favoring Tweedledum's views over Tweedledee's. Alice is quite correct to reprimand me for this. I like my theories to be as simple as possible, and those theories which can explain the largest number of independent facts with the fewest assumptions are, in my view, to be preferred. Tweedledum's views do add up to some sort of self-consistent story. However, I fully realize that this does not mean that the picture is correct. Tweedledee could equally well be correct at this stage. When I say I prefer a particular picture, I am simply describing the preferred framework I have adopted for the pursuit of my own research goals.

The key point is that nowadays we can at least ask these questions and have realistic hopes of resolving some of them by astronomical observation. As I have suggested, the Space Telescope will have a special role to play in these studies.

9. *Another Old Friend*

1.   The White Knight meets Alice in chapter 8 of *Through the Looking-glass.* Being based upon a chess piece, he keeps falling off his horse because the Knight moves lopsidedly on the chess board. As Martin Gardner points out in his annotated edition of the Alice books, many scholars have surmised that the White Knight is a caricature of Lewis Carroll (Charles Dodgson) himself. He is certainly the kindest of all the characters in the Alice stories and he is clearly fond of Alice. He seems the right character to give Alice the best advice about the importance of the Space Telescope.

2.   The three examples given by the White Knight illustrate how discoveries made in astronomy can come as a complete surprise to theoreticians. For example, it was a great surprise when the pulsating radio sources observed at long radio wavelengths turned out to be neutron stars. No one had thought of looking for neutron stars in this way because it was not realized that very intense pulses of radio emission would be produced by magnetized rotating neutron stars. Similarly, no theoretician would have predicted that there would be intense jets of very high energy particles emitted by the nuclei of radio galaxies and quasars. Finally, the binary neutron star provides a perfect clock in a rotating frame of reference, which is ideal for making subtle tests of general relativity.

APPENDIX 1

The Specification and Performance Goals of the Hubble Space Telescope

Telescope aperture	2.4 meters
System focal ratio	f/24
Optical design	Ritchey-Chrétien Cassegrain
Total obscured area	13.8% aperture area
Total field of view	28 arcmin diameter
Spectral range	115 nm to 1 mm
Optical performance: system wavefront error	$\lambda/13.5$ at 632.8 nm
Radius of 70% encircled energy from a point source at 633 nm	0.1 arcsec
Faint object sensitivity subject to viewing in directions greater than (1) 50° from Sun, (2) 70° from Earth's limb, and (3) 15° from the Moon.	Point objects having $m = 27$ or brighter; extended sources to a limit of 23 mag arcsec^{-2} in visual waveband. Both observations to be made in 10 hours of observation.
Pointing stability	0.007 arcsec on a guide star having $m_v = 14$ to 14.5 in the wavelength range 400–800 nm.
Minimum throughput to focal plane	60% at 632.8 nm 40% at 120 nm

The Satellite

Launch weight	11,600 kilogram (25,500 lb or 12.75 tons)
Satellite length	13.1 meters (43.5 ft)
Satellite diameter	4.27 meters (14 ft)
Size of Solar Panels	Each 11.8×2.3 meter (39.4×7.8 ft)
Power supplied by Solar Array	Minimum of 2400 watts

Satellite Orbit

Circular Earth orbit serviceable by the Space Shuttle.

Altitude	about 600 km
Inclination	28.5°

APPENDIX 1

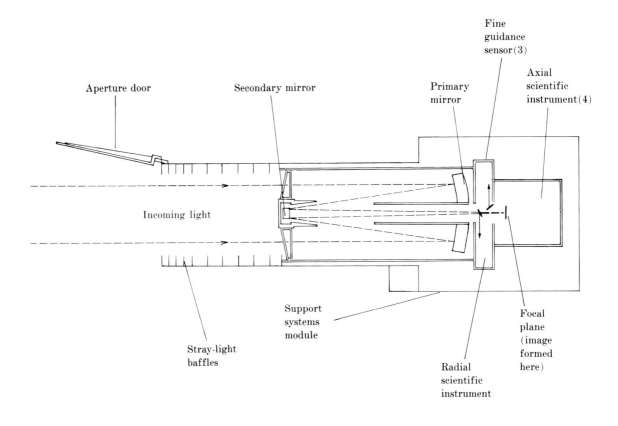

Aperture door

Secondary mirror

Primary mirror

Fine guidance sensor(3)

Axial scientific instrument(4)

Incoming light

Support systems module

Stray-light baffles

Radial scientific instrument

Focal plane (image formed here)

[188]

APPENDIX 2

The Specifications and Expected Performance of the Scientific Instruments for the Hubble Space Telescope

THE WIDE FIELD/PLANETARY CAMERA

The camera can operate with two different focal ratios, f/12.9 or f/30, the former being known as the wide-field mode and the latter as the planetary mode. In each mode the detector consists of an array of four thinned, back-illuminated CCDs (Charge Coupled Devices), each consisting of 800×800 pixels.

	Wide Field Mode	Planetary Mode
Focal ratio	f/12.9	f/30
Picture format	1600×1600 pixels	1600×1600 pixels
Field of view	2.57×2.57 (arcmin)^2	66.7×66.7 (arcsec)^2
Pixel size	0.1 arcsec	0.043 arcsec
Wavelength range	115–1100 nm	115–1100 nm
Dynamic range (per pixel, single exposure)	460*	460*
Maximum S/N (per pixel, single exposure)	170	170
Overall dynamic range for stars (V)	about 9 to 29	about 8 to 29
Exposure times	Minimum exposure time is 0.11 seconds Maximum length of target integrations is 100,000 seconds. In practice, the exposure times will be limited by saturation effects and cosmic ray events.	

The camera contains 12 filter wheels, each with 4 filters and a clear "home" position. The filters include a wide range of narrow- and broad-line filters as well as polarization filters and transmission gratings.

*The full-well capacity of the CCD detectors is 30,000 electrons and the r.m.s. read-out noise per pixel is 13 ± 2 electrons. The quoted dynamic range is the full-well capacity divided by five times the read-out noise.

For more details, see *Wide Field/Planetary Camera Instrument Handbook*, by R. Griffiths, STScI publications.

FAINT-OBJECT SPECTROGRAPH

The Faint-Object Spectrograph employs two independent optical channels, which image the dispersed image onto two digicon detectors. One detector (the blue digicon) is sensitive to radiation in the wavelength range 110 to 550 nm and the other (the red digicon) in the wavelength range 180 to 850 nm.

Spectral range	110–850 nm
Spectral resolving power	(1) $R = \lambda/\Delta\lambda = 1300$ between 110 and 850 nm using a filter grating wheel to allow selection of six slightly overlapping spectral regions.
	(2) $R = \lambda/\Delta\lambda = 250$ between 110 and 850 in three spectral bands covered by two gratings and a prism.
Detectors	512 diode linear array digicons
Aperture sizes	A wide range of observing apertures is available, including round apertures of diameters 0.3, 0.5, and 1 arcsec as well as pairs of square apertures of sizes 0.1, 0.25, 0.5, and 1.0 arcsec for observing object and sky simultaneously. Some larger apertures are available for extended objects and target acquisition.
Dynamic range	At least 2.5×10^7
Exposure times	Minimum time for time resolved spectroscopy $= 50$ msec
	No limit on maximum length of target integration

For more details, see *Faint-Object Spectrograph Instrument Handbook*, by Holland C. Ford, STScI publications.

FAINT-OBJECT CAMERA

The Faint-Object Camera is designed to exploit the ultimate in resolution capabilities of the Hubble Space Telescope. The observing modes of the camera are fourfold: (1) imaging in the f/96 mode, (2) imaging in an f/48 mode, (3) imaging in an f/288 mode, and (4) long-slit spectroscopy in the f/48 mode.

	f/48	f/96	f/288
Field of view (arcsec)^2	44×44	22×22	7.6×7.6
Pixel sizes (arcsec)^2	0.044×0.088	0.022×0.044	0.0075×0.015

Smaller fields can be imaged at higher spatial resolution and extended dynamic range.

Angular resolution	Diffraction-limited performance at all wavelengths. At the shortest ultraviolet wavelengths, the Faint-Object Camera will obtain diffraction-limited performance by image reconstruction techniques and should result in angular resolution of about 0.01 arcsec.
Wavelength range	115–700 nm
Dynamic range	$m_v = 21$ to 30 on point sources; 15 to 22 visual mag (arcsec)$^{-2}$ for extended sources
Number of pixels	Normally 512×512 pixels for 16-bit data words, but 512×1024 available for 8-bit data storage.

| Exposure time | The detector is a photon-counting system and thus the lower limit of time resolution depends upon the rate at which photons can be separately detected. No limit on maximum length of exposure time. |

For more details, see *Faint-Object Camera Handbook*, by F. Paresce, STScI publications.

HIGH-RESOLUTION SPECTROGRAPH

The High-Resolution Spectrograph employs two digicon detectors, one of them sensitive to the wavelength range 105–170 nm, the other to the range 115–320 nm.

Spectral resolving power $R = \lambda/\Delta\lambda$	2000	2×10^4	10^5
Spectral range (nm)	105–180	105–320	105–320

Although the nominal short-wavelength limit of the Hubble Space Telescope is 115 nm, there will be some sensitivity at shorter wavelengths. In particular, lines of deuterium and other heavy elements will be observable in the waveband 105–120 nm.

6 gratings are mounted on a rotating carousel.

| Detectors | 512 diode linear array digicons. |

Thus, for example, in the 2×10^4 resolution mode the plate scale at the detector is about 0.008 nm per pixel, and so approximately 4 nm of the available spectral range are recorded by a single observation with the 500 science diodes of the arrays.

Aperture sizes (arcsec^2)	2×2	0.25×0.25
Dynamic range	At least 10^7	
Exposure times	Minimum time for time resolved spectroscopy = 50 msec.	

No limit on maximum length of target integration.

For more details, see *High-Resolution Spectrograph Instrument Handbook*, by Dennis Ebbets, STScI publications.

HIGH-SPEED PHOTOMETER/POLARIMETER

The High-Speed Photometer/Polarimeter consists of four image dissector tubes and one photomultiplier tube. Three of the image dissectors are used for photometry and the fourth for polarimetry. The photomultiplier tube is used for simultaneous two-color photometry with one of the image dissector tubes used for photometry.

Spectral range	22 UV and visual filters spanning the wavelength range 120 to 700 nm for photometry.
	4 UV filters in the wavelength range 200–330 nm for polarimetry.
Apertures	0.4 and 1.0 arcsec diameter.
Photometric accuracy	0.1% from $V = 0$ to $V = 20$.
Polarimetric accuracy	0.1%
Time resolution	10μs in pulse-counting mode for count rates less than 10^6 counts s^{-1}.
	~1 ms in current mode for count rates greater than 10^6 counts s^{-1}.

There are no moving parts in this instrument, the different filters and apertures being selected by slewing the Hubble Space Telescope.

For more details, see *High-Speed Photometer Instrument Handbook*, by Richard L. White, STScI publications.

FINE GUIDANCE SENSORS

The Fine Guidance Sensors are primarily used for locking the Hubble Space Telescope onto its scientific targets with very high accuracy. The specification requires the Fine Guidance Sensors to maintain the targets within the field of view of the instruments with the accuracy of 0.007 arcsec. In addition, the sensors can be used to measure the relative positions of stars within the field of view of the sensors with very high accuracy.

Accuracy of measurement of relative separations of three stars
brighter`than mag. 17 within a 20 square arcmin field of view :   0.0016 arcsec (r.m.s.)

This accuracy is expected to be obtained for observations taken within an observing time of less than about 20 minutes.

Time to reach 17 magnitude for stars :   about 2 minutes

Magnitude range of stars which can be measured :   4 to 17

The field of view available for astrometric studies is defined by the 3 pickle-shaped segments of the Fine Guidance Sensors. As shown in the figure, the three segments extend radially from 10.2 arcmin to 14 arcmin and azimuthally for nearly a 90° arc. The total area of sky included in the segments is 69 arcmin^2.

For more details, see *Fine Guidance Sensor Instrument Handbook*, by Alain Fresneau, STScI publications.

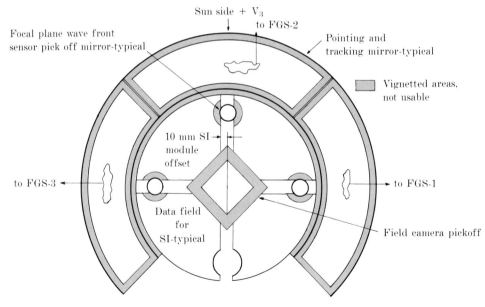

The Focal Plane arrangement

Bibliography

There is a vast literature on astronomy and cosmology at the popular and the technical levels. I give here a list of some introductory texts I have recommended to enthusiasts for further reading.

GENERAL REFERENCES

Abell, G. O. *Exploration of the Universe*, 4th ed. Philadelphia : Saunders College Publishing, 1982.

Abell, G. O. *The Realm of the Universe*, 3d ed. Philadelphia : Saunders College Publishing, 1984.

The Cambridge Encyclopedia of Astronomy. Edited by S. Milton. London : Jonathan Cape, 1977.

The Cambridge Atlas of Astronomy. Edited by J. Audouze and G. Israel. Cambridge : Cambridge University Press, 1985.

Henbest N., and M. Marten. *The New Astronomy*. Cambridge : Cambridge University Press, 1983.

Shu, F. H. *The Physical Universe: An Introduction to Astronomy*. Mill Valley, Calif. : University Science Books, 1982.

Kaufman III, W. J. *Universe*. New York : W. H. Freeman and Co., 1985.

SOLAR SYSTEM

The New Solar System. Edited by J. Kelly Beattie, B. O'Leary, and A. Chaiken. Cambridge : Cambridge University Press, 1981.

Briggs, G., and F. Taylor. *The Cambridge Photographic Atlas of the Planets*. Cambridge : Cambridge University Press, 1982.

Brandt, J. C., and R. D. Chapman. *Introduction to Comets*. Cambridge : Cambridge University Press, 1981.

STARS AND STELLAR EVOLUTION

R. Kippenhahn. *One Hundred Billion Suns*. London : Weidenfeld and Nicolson, 1983.

R. J. Taylor. *Stars: Their Structure and Evolution*. London : Wykeham Publications, 1970.

I. S. Shklovskii. *Stars: Their Birth, Life, and Death*. Translated by R. B. Rodman. New York : W. H. Freeman and Co., 1978.

INTERSTELLAR MEDIUM

Spitzer, L., Jr. *Searching between the Stars*. New Haven : Yale University Press, 1982.

GALAXIES

All the general references above contain good descriptions of galaxies.
Tayler, R. J. *Galaxies: Structure and Evolution.* London: Wykeham Publications, 1978.
Michalas, D., and J. Binney, *Galactic Astronomy: Structure and Kinematics.* New York: W. H. Freeman and Co., 1981. *(Quite technical.)*

ACTIVE GALAXIES AND QUASARS

There is no simple introductory text devoted to active galaxies and quasars, but there are good chapters on these objects in all the general references above.
Shapiro, S. L., and S. A. Teukolsky. *Black Holes, White Dwarfs, and Neutron Stars: The Physics of Compact Objects.* New York: Wiley-Interscience, 1983. *(An excellent technical book.)*

COSMOLOGY

Harrison, E. R. *Cosmology: The Science of the Universe.* Cambridge: Cambridge University Press, 1981.
Weinberg, S. *The First Three Minutes: A Modern View of the Origin of the Universe.* New York: Basic Books, 1976. London: Andre Deutsch, 1977.
Barrow, J. D., and J. Silk. *The Left Hand of Creation: The Origin and Evolution of the Expanding Universe.* New York: Basic Books, 1983. London: Heinemann, 1984.
J. Silk. *The Big Bang: The Creation and Evolution of the Universe.* New York: W. H. Freeman and Co., 1980.

ASTRONOMICAL DICTIONARIES

The following dictionaries may be useful for definitions of unfamiliar terms.
V. Illingworth. *The Macmillan Dictionary of Astronomy.* London: Macmillan Reference Publications, 1979.
J. Hopkins. *A Glossary of Astronomy and Astrophysics.* Chicago: University of Chicago Press, 1976.
J. Mitton. *Key Definitions in Astronomy.* London: Frederick Muller, 1980.

ASTRONOMICAL AND PHYSICAL DATA

C. W. Allen. *Astrophysical Quanties.* 3d ed. London: Athlone Press, 1973.
Kaye G. W. C., and T. H. Laby. *Tables of Physical and Chemical Constants.* London: Longman Group, 1973.

Index

The principal references in the Annotations to topics of special importance are indicated in **boldface**.

Picture Credits

Sir John Tenniel: pages 3, 4, 7, 8, 11, 12, 13, 15, 25, 27, 28, 34, 35, 39, 49, 52, 53, 59, 70, 71, 89, 91

NASA: pages 9, 10, 16, 18, 20, 42 *bottom*, 92, 101, 103, 104, 107, 109, 156, 188, 190

NASA /Jet Propulsion Laboratory: pages 17, 108

Official U.S. Navy photograph: page 19

Sky Publishing Corp.: page 21

From "The Space Telescope," by John N. Bahcall and Lyman Spitzer, Jr., copyright 1982, Scientific American, Inc., All rights reserved: page 29

Copyright Royal Observatory, Edinburgh: pages 22, 31, 37, 38, 40, 45 *bottom*, 46 *bottom*, 47, 50, 51, 79, 117 *bottom*, 122, 138

European Space Agency: page 32

The Computer Book, by Robin Bradbeer, Peter de Bono and Peter Laurie, copyright British Broadcasting Corp., 1982: page 36

Copyright Ango-Australian Telescope Board, 1983: page 42 *top*

Drawings by Stephen Kraft: pages 43, 60, 63, 74, 78, 80, 81, 84 *bottom*, 88, 98, 99, 111, 114, 115, 117 *top*, 119, 120, 121, 123, 124, 129, 133, 135, 136, 137, 140, 142, 146, 148, 154, 157, 158, 162, 166, 167, 168, 169, 170, 171, 174, 176, 177, 181, 183

Palomar and Mount Wilson Observatories: pages 45 *top*, 46 *top*, 66, 118 *bottom*

John L. Tonry, Institute for Advanced Study: page 48

James E. Gunn: page 56

Copyrights Royal Greenwich Observatory: page 62

Halton C. Arp: page 64

After A. R. Sandage, *Observatory* 88 (1968), 99: page 73

M. S. Longair, after data listed in *American Institute of Physics Handbook*, 2d ed., 1963, pp. 6-224–6-227: page 74

F. Macchetto: page 102

Reprinted courtesy of R. A. Perley and the *Astrophysical Journal*, published by the University of Chicago Press; copyright 1984 The American Astronomical Society: page 65

Reprinted by permission from *Nature* 290, pp. 365–68, copyright 1981 Macmillan Journals Ltd: page 67

Journal of the Royal Astronomical Society 13 (1972), 202: page 72

Pontificia Academia Scientarium: pages 75, 83, 88

From M. Seldner, B. Siebers, E. J. Groth and P. J. E. Peebles, *Astronomical Journal* 82 (1977): page 82

Monthly Notices of the Royal Astronomical Society 198 (1982), 194; R. F. Carswell: page 84

Monthly Notices of the Royal Astronomical Society 211 (1984), 849: page 86

Monthly Notices of the Royal Astronomical Society 160 (1972), 1: page 132

Monthly Notices of the Royal Astronomical Society 209 (1984), 159: page 175

Davis Meltzer: page 116

Courtesy of Mullard Radio Astronomy Observatory: page 118 *top*

John D. Kraus, from *Radio Astronomy:* page 128

Harvard College Observatory; C. G. Wynn-Williams: page 130

Courtesy of Fred Seward, Harvard-Smithsonian Center for Astrophysics: page 131

M. and T. Keskula, Lund Observatory: page 139

John N. Bahcall and R. Soneira: page 144

E. von Groningen: page 150

V. M. Lyutii: page 151

Reprinted courtesy of Paul L. Richards and the *Astrophysical Journal*, published by the University of Chicago Press; copyright The American Astronomical Society: page 163

Reprinted courtesy of R. V. Wagoner and the *Astrophysical Journal*, published by the University of Chicago Press; copyright The American Astronomical Society: page 165

Michael Seldner: page 179

R. J. Weymann and C. Foltz: page 182